AF688634

PACKAGING

BY SANJIVAN SAINI

While every precaution has been taken in the preparation of this book, the publisher assumes no responsibility for errors or omissions, or for damages resulting from the use of the information contained herein.
PACKAGING
First edition. September 20, 2023.
Copyright © 2023 SANJIVAN SAINI.
Written by SANJIVAN SAINI.

Over View :

"Packaging" by Sanjivan Saini is a comprehensive book that provides an overview of various aspects of packaging. It covers a wide range of topics related to packaging, including special packaging, safety precautions, exportable goods packaging, dangerous goods packaging, packaging for cold products, freezing products, dairy products, and packaging for hot products. Additionally, the book delves into the cost involved in packaging, its components, and various other relevant subjects. Here is a brief overview of the main units covered in the book:

1. Packaging: This unit likely introduces the fundamentals of packaging, including its purpose, importance, and various applications across industries. It may cover topics such as packaging materials, design, and functions.

2. Special Packaging: The section on special packaging explores specific packaging requirements for products with unique characteristics or needs, such as medical devices, fragile items, or hazardous substances.

3. Safety Precautions: Safety is paramount in packaging. This unit likely discusses safety measures and guidelines for packaging, including how to handle hazardous materials safely and ensure compliance with regulations.

4. Exportable Goods Packaging: Exporting goods requires careful packaging to withstand the rigors of transportation. This unit would cover the packaging considerations for goods destined for international markets.

5. Dangerous Goods Packaging: This unit delves into the packaging requirements for dangerous goods, including hazardous chemicals, radioactive materials, and other substances that require specialized containment.

6. Packaging for Cold Products, Freezing Products, Dairy Products: These sections likely focus on packaging solutions for perishable goods, such as food items that need refrigeration or freezing to maintain their quality and safety.

7. Packaging for Hot Products: In contrast to the previous unit, this section would cover packaging designed to handle products that require high-temperature resistance, such as hot food items or industrial materials.

8. Cost Involved, Component of Cost Involved: This unit would explore the economic aspects of packaging, including the various costs associated with packaging materials, design, production, and transportation.

The book "Packaging" by Sanjivan Saini is likely to be a valuable resource for students, professionals, and anyone interested in gaining a comprehensive understanding of the diverse aspects of packaging. It covers a wide range of topics, providing practical insights into designing effective and safe packaging solutions for different types of products and industries.

Table of Contents

UNIT-1. Packaging — *1*

1.1 Packaging — 1

1.2 Nature of Packaging — 3

1.3 Various materials used — 6

1.4 Reusable materials used — 8

1.5 Type of Packaging — 11

1.6 Cost Involved, Component of Cost Involved — 13

1.7 Economy in Packaging — 16

1.8 Freight Vs Packaging — 18

Packaging: — 19

Freight: — 19

1.9 Material Handling Vs Packaging And Weight Constraint — 21

Material Handling: — 21

Packaging and Weight Constraint: — 22

1.10 Consumer Attracting Packaging — 24

1.11 Packaging for safety — 26

1.12 Packaging to boost sale — 29

1.13 Strengthening Methods — 31

1.14 Packaging Functions — 34

1.15 Benefits And Challenges — 36

Benefits of Packaging: — 37

Challenges of Packaging: — 38

UNIT-2. Special Packaging — *44*

2.1 Packaging for special product — 44

2.2 Packaging For Cold Product, Freezing Product, Dairy Product — 47

Packaging for Cold Products (Chilled Products): — 47

Packaging for Freezing Products (Frozen Products): — 48

Packaging for Dairy Products: — 49

2.3 Packaging for hot products 50

2.4 Fruits /vegetable Packaging 52

2.5 Packaging of grains 55

2.6 Packaging of pharmacy products 57

2.7 Packaging of food products 60

2.8 Packaging Of Liquids, Water, Edible Oil 62

Water Packaging: 62

Edible Oil Packaging: 63

2.9 Packaging Of Cosmetics And Powdered Articles 65

Cosmetics Packaging: 65

Powdered Articles Packaging: 66

2.10 Packaging Guide Lines 67

2.11 Single box and box in box Packaging –storing in cold / frozen storage 70

Single Box Packaging: 70

Box-in-Box Packaging: 71

UNIT-3. Safety precautions 75

3.1 Safety precautions in Packaging 75

3.2 Difficulty in Packaging 78

3.3 Defects in Packaging 81

3.4 Label Mixing, Label Error 83

Label Mixing: 83

Label Error: 84

3.5 Lot No, Missing Lot No 85

3.6 Date of Packaging &Manufacture 88

Date of Packaging: 88

Date of Manufacture: 88

3.7 Storage Instruction 89

3.8 Supervision And Examination 91

Supervision: 92

Examination: 92

3.9 Investigation on Discrepancies And Tamper Evident 94

Investigation on Discrepancies: 94
Tamper-Evident Features: 94

3.10 Packaging for shipment 96

3.11 Packaging For Air Cargo, Rail Transport 98
Packaging for Air Cargo: 98
Packaging for Rail Transport: 100

3.12 Requirements sensitivity of Packaging for air cargo 101

3.13 Performance Oriented Packaging 103

3.14 Warehouse Labeling 106

3.15 Hygroscopic protection to be taken care 109

3.16 IATA regulations and lashing 111
IATA Regulations: 111
Lashing: 112

3.17 Securing in air transport 113

UNIT-4. Exportable goods Packaging 120

4.1 Introduction: 120

4.2 Packaging for export 123

4.3 Precaution In Packaging 126

4.4 Quality Packaging 128
Characteristics of Quality Packaging: 129
Benefits of Quality Packaging: 130

4.5 Country's Labeling, Country's Language 131

4.6 Colour Restrictions, Provision Country of Origin 133
Color Restrictions: 133
Provision of Country of Origin: 134

4.7 ISO standard 135
Key characteristics of ISO standards include: 135

4.8 Physical Protection, Quality Protection 137
Physical Protection: 137
Quality Protection: 139

4.9 Product Information, Machinability 140

Product Information: 140
Machinability: 141

4.10 Storage And Transport 143

Storage: 143
Transport: 144

4.11 Recyclability-Ventilation of Package 145

Recyclability: 146
Ventilation of Package: 147

4.12 Cushioning Materials And Moisture Absorbing System 148

Cushioning Materials: 148
Moisture Absorbing System: 150

UNIT-5. Dangerous goods Packaging 156

5.1 Packaging of dangerous goods 156

5.2 Packaging of Chemicals, Oils, Toxic Gasses, LPG,LNG. 159

Chemicals: 159
Oils: 160
Toxic Gases: 160
LPG (Liquefied Petroleum Gas) and LNG (Liquefied Natural Gas): 161

5.3 Can, Drum, Cylinder Packaging 161

Cans: 162
Drums: 162
Cylinders: 162

5.4 Portable Tanker Packaging, Grating of Tanker 163

Portable Tanker Packaging: 163
Grating of Tanker: 164

5.5 Packaging of radioactive materials 166

5.6 Inner Container, Outer Container 168

Inner Container: 168
Outer Container (Overpack): 169

5.6 Packaging groups I,II,III 171

Packaging Group I: 171
Packaging Group II: 171
Packaging Group III: 172

5.7 Pressure receptacles for liquids and solids 172

Pressure Receptacles for Liquids: 173
Pressure Receptacles for Solids: 173

5.8 Packaging instruction of UN3373, UN3245 175
Packaging Instruction for UN3373 - Biological Substance, Category B: 175
Packaging Instruction for UN3245 - Genetically Modified Organisms (GMOs): 177

5.9 Labelling of dangerous goods, signs /symbols to be used 177

5.10 IATA regulations for dangerous goods 180

UNIT-1. Packaging

LESSON OUTLINE-

- Packaging Need, Necessity,
- Nature Of Packaging, Various Materials Used,
- Reusable Materials Used And
- Type Of Packaging
- Cost Involved, Component Of Cost Involved,
- Economy In Packaging, Freight Vs Packaging,
- Material Handling Vs Packaging And Weight Constraint
- Consumer Attracting Packaging,
- Packaging For Safety,
- Packaging To Boost Sale,
- Strengthening Methods And Packaging Functions.

1.1 Packaging

Packaging refers to the process of enclosing or protecting products or goods to ensure their safe storage, handling, transportation, and distribution. It involves designing and manufacturing containers, boxes, wrappers, or other materials that surround a product and provide necessary information about the product's contents. Packaging serves multiple purposes, making it essential in various industries:

1. **Protection:** One of the primary purposes of packaging is to safeguard the product from damage,

contamination, or breakage during transit or storage. It helps preserve the product's quality and ensures it reaches the consumer in good condition.

2. **Safety:** Packaging plays a crucial role in protecting consumers from potential hazards associated with certain products. For example, child-resistant packaging for medicines or safety seals on food items prevent accidental consumption.

3. **Identification and Information:** Packaging provides essential information about the product, such as its name, ingredients, usage instructions, manufacturing date, expiry date, and nutritional values. This information helps consumers make informed choices and also aids regulatory compliance.

4. **Convenience:** Packaging is designed to make it easy for consumers to handle, use, and store the product. Features like resealable pouches, easy-pour spouts, or single-serving packs enhance convenience.

5. **Branding and Marketing:** Packaging serves as a powerful marketing tool to attract customers. Unique and visually appealing packaging can help differentiate a product from its competitors and create a memorable brand image.

6. **Transportation and Logistics:** Packaging facilitates the efficient movement of goods through the supply chain. It ensures easy stacking, handling, and storage

of products during transportation, reducing the risk of damage.

7. **Extended Shelf Life:** Certain packaging materials and techniques can extend the shelf life of perishable products by protecting them from external factors like moisture, light, or air.

8. **Compliance and Regulations:** Different industries have specific packaging regulations and requirements set by governments and regulatory bodies to ensure safety, hygiene, and environmental protection.

In summary, packaging is a necessity for businesses and consumers alike. It serves not only as a protective measure for products but also as a means of communication and marketing. By addressing various practical and regulatory needs, packaging ensures that products can reach consumers in a safe and appealing manner, contributing to overall consumer satisfaction and brand success.

1.2 Nature of Packaging

The nature of packaging is multifaceted and dynamic, reflecting its role as a critical aspect of modern-day product distribution and consumption. Here are some key aspects that define the nature of packaging:

1. **Functionality**: Packaging serves a practical purpose of protecting the product during storage, transportation, and handling. It ensures that the product reaches the end consumer in optimal

condition. The choice of packaging material and design is based on the specific requirements of the product, considering factors like fragility, perishability, and sensitivity to environmental conditions.

2. **Innovation**: The packaging industry continuously evolves with advancements in technology, materials, and design. Manufacturers and businesses often invest in research and development to create innovative packaging solutions that offer better protection, convenience, and sustainability.

3. **Sustainability**: In recent years, there has been a growing focus on sustainable packaging due to environmental concerns. The nature of packaging is shifting towards eco-friendly materials, reduced waste, and recyclability or biodegradability. Sustainable packaging aims to minimize the impact on the environment while maintaining its functional role.

4. **Aesthetics and Branding**: Packaging is an essential part of a product's branding and marketing strategy. Eye-catching designs, attractive colors, and logos help create brand recognition and influence consumer purchasing decisions.

5. **Regulatory Compliance**: The packaging industry is subject to various regulations and standards set by governments and industry bodies. These regulations may pertain to safety, labeling, food contact materials, and environmental impact.

6. **Cost-Effectiveness**: While packaging needs to be functional and visually appealing, it must also be cost-effective for businesses. Balancing the cost of packaging with its benefits is crucial in product pricing and overall profitability.

7. **Consumer-Focused**: Packaging is designed with consumers in mind. It aims to make products easy to use, handle, and store, enhancing the overall customer experience. Features like resealable packaging, portion control, and convenience packs cater to consumer preferences.

8. **Customization**: Different products require unique packaging solutions. The nature of packaging allows for customization to meet specific product needs and target different consumer demographics.

9. **Globalization**: With the growth of international trade, packaging must comply with various international standards and adapt to diverse cultural preferences and regulations across different markets.

10. **Technological Integration**: Advancements in technology have allowed for smart packaging solutions, incorporating features like RFID tags, temperature sensors, and interactive elements that provide real-time information and enhance product functionality.

In conclusion, the nature of packaging is a dynamic blend of functionality, innovation, sustainability, branding, and compliance. It plays a vital role in protecting products, engaging consumers, and contributing to the overall success of businesses in a rapidly changing global market.

1.3 Various materials used

Packaging materials come in a wide range of options, each chosen based on the specific requirements of the product, its intended use, and environmental considerations. Here are some of the various materials commonly used in packaging:

1. **Paper and Cardboard**: Paper and cardboard are widely used for packaging due to their versatility, recyclability, and cost-effectiveness. They are commonly used for boxes, cartons, bags, and labels.

2. **Plastics**: Plastics are prevalent in packaging due to their durability, flexibility, and barrier properties. Common plastic packaging materials include polyethylene (PE), polypropylene (PP), polyethylene terephthalate (PET), and polystyrene (PS). However, there's increasing concern about plastic waste and its environmental impact.

3. **Glass**: Glass is commonly used for packaging beverages, sauces, and cosmetics. It is non-reactive and offers an excellent barrier against moisture and oxygen. Glass is recyclable and retains its quality through multiple recycling loops.

4. **Metals**: Metals like aluminum and steel are used for packaging canned beverages, foods, and aerosols. They provide excellent barrier properties, protect against light and air, and are easily recyclable.

5. **Biodegradable and Compostable Materials**: As part of the growing trend towards sustainable packaging, biodegradable and compostable materials made from sources like cornstarch, sugarcane, or algae are gaining popularity. They break down naturally, reducing environmental impact.

6. **Wood**: Wood is used for packaging larger or heavier items, such as wooden crates or pallets, which are commonly used for industrial and bulk transportation.

7. **Fabrics and Textiles**: Cloth or fabric packaging is utilized for products like apparel, gifts, and specialty items, offering a unique and reusable packaging solution.

8. **Foam**: Foam materials, such as expanded polystyrene (EPS) or polyethylene foam, are used for cushioning and protecting fragile items during transit.

9. **Flexible Films**: Flexible films made from materials like polyethylene, polypropylene, or laminates are used for packaging snacks, frozen foods, and various consumer goods. They offer lightweight and cost-effective packaging solutions.

10. **Ceramics**: Ceramics are used for specialized packaging of delicate items, such as electronics or high-end gifts.

11. **Vacuum Packaging**: This involves removing air from the packaging to extend the shelf life of products and reduce spoilage.

12. **Active and Intelligent Packaging**: These are innovative packaging types that can interact with the product or the environment to enhance product freshness or provide real-time information to consumers.

It's worth noting that the choice of packaging material is influenced by factors such as the product's nature, transportation requirements, shelf life, environmental impact, and consumer preferences. As sustainability becomes more critical, there's an increasing focus on developing and using materials that minimize waste and have a reduced environmental footprint.

1.4 Reusable materials used

Reusable materials are gaining popularity in packaging as part of the broader sustainability movement, aimed at reducing waste and minimizing the environmental impact of packaging. These materials are designed to be used multiple times, either for the same product or repurposed for other applications. Here are some common reusable materials used in packaging:

1. **Glass**: Glass containers and bottles are a classic example of reusable packaging. They can be easily cleaned and sanitized for refilling with various beverages, sauces, or other products.
2. **Metal Containers**: Metal containers, such as stainless steel or aluminum bottles and cans, are also reusable. They are often used for beverages, personal care products, and food items.
3. **Fabric Bags**: Reusable fabric bags made from materials like cotton, canvas, or jute are a sustainable alternative to single-use plastic bags. They can be used for grocery shopping, retail purchases, or as promotional items.
4. **Silicone**: Silicone is a flexible and durable material that is commonly used for reusable food storage bags, food wraps, and other kitchen containers.
5. **Bamboo**: Bamboo is a fast-growing and renewable resource used for various reusable items like bamboo straws, cutlery, and food containers.
6. **Sturdy Plastic Containers**: While single-use plastics are a concern, some thicker and more durable plastic containers are designed for multiple uses, such as food storage containers and bins.
7. **Stainless Steel Containers**: Stainless steel containers are durable, hygienic, and often used as reusable lunchboxes or food storage containers.

8. **Wooden Crates**: Wooden crates are reusable and commonly used for shipping and storing various goods.

9. **Tote Bags**: Reusable tote bags made from cloth, jute, or other materials are widely used as an eco-friendly alternative to plastic bags.

10. **Mason Jars**: Mason jars are versatile and can be reused for food storage, canning, crafting, and even as drinking glasses.

11. **Mesh Produce Bags**: These lightweight and breathable bags are used to replace single-use plastic bags for purchasing fruits and vegetables.

12. **Wax Wraps**: Beeswax wraps are an eco-friendly alternative to plastic wrap and can be washed and reused multiple times.

13. **Reusability Kits**: Some companies offer reusable packaging solutions, where customers can return empty containers to be refilled or exchanged for new products.

Promoting the use of reusable packaging materials helps reduce the amount of waste generated and contributes to a more sustainable and environmentally responsible approach to product distribution and consumption. Additionally, reusable packaging can also be economically beneficial in the long run, both for businesses and consumers.

1.5 Type of Packaging

Packaging comes in various types, each designed to cater to specific product requirements, industry standards, and consumer preferences. Here are some common types of packaging:

1. **Boxes and Cartons**: Boxes and cartons are among the most widely used packaging types. They come in various sizes and shapes and are commonly used for retail packaging, shipping, and storage.

2. **Bottles and Jars**: Bottles and jars are used to package liquids, beverages, condiments, cosmetics, and pharmaceutical products. They are available in different materials like glass, plastic, or metal.

3. **Bags and Pouches**: Bags and pouches are flexible packaging options used for snacks, dry goods, and liquids. They often have resealable closures for convenience.

4. **Cans and Drums**: Cans are commonly used to package beverages and canned foods, while drums are larger containers used for industrial and bulk packaging.

5. **Blister Packs**: Blister packs consist of a plastic pocket or cavity that holds the product and a paper or cardboard backing. They are commonly used for pharmaceuticals, small consumer goods, and electronics.

6. **Clamshells**: Clamshell packaging consists of two hinged plastic halves that encase the product. They are frequently used for electronics, toys, and small hardware items.
7. **Tubes**: Tubes are used for packaging products like toothpaste, lotions, and creams. They are typically made of plastic or aluminum.
8. **Wraps and Films**: Wraps and films are thin, flexible materials used for wrapping and protecting various products, including food items and consumer goods.
9. **Labels and Sleeves**: Labels provide essential information and branding on product packaging. Sleeves are full or partial coverings that fit around a product's primary packaging.
10. **Pallets and Crates**: Pallets are used for bulk transportation and storage of goods, while crates provide sturdy protection for heavy or fragile items.
11. **Tins and Tubs**: Tins and tubs are used for packaging food items like cookies, candies, and confectionery.
12. **Ampoules and Vials**: These small, sealed containers are commonly used to store pharmaceuticals, chemicals, and liquid samples.
13. **Sachets and Stick Packs**: These single-use packages hold small quantities of products like condiments, instant coffee, and powdered beverages.

14. **Shrink Wraps**: Shrink wraps are thin plastic films that conform tightly to the product when heated, providing protection and tamper-evidence.

15. **Composite Cans**: Composite cans are cylindrical containers made from cardboard and lined with a protective layer, used for packaging snacks, chips, and powdered goods.

16. **Tetra Paks**: Tetra Paks are used to package liquid products like milk, juices, and dairy alternatives. They consist of laminated layers of paper, aluminum, and plastic.

17. **Gift Boxes**: Gift boxes are decorative packaging used for presenting gifts and specialty items.

The type of packaging chosen depends on various factors, such as the nature of the product, its size, weight, fragility, and the intended distribution and sales channels. Additionally, businesses also consider the environmental impact and consumer preferences when selecting the appropriate packaging type.

1.6 Cost Involved, Component of Cost Involved

The cost involved in packaging can vary significantly depending on factors such as the type of packaging, the material used, the volume of production, and the complexity of the design. The components of packaging costs can be broadly categorized as follows:

1. **Material Costs**: This includes the cost of the actual packaging materials used, such as paper, cardboard, plastic, glass, metal, or any other material. The cost will depend on the quantity and quality of the material used.

2. **Labor Costs**: Labor costs include expenses related to the workforce involved in packaging, including design, printing, assembly, and quality control. Highly specialized or intricate packaging may require skilled labor, impacting the overall cost.

3. **Design and Printing**: Packaging design and printing costs cover the creation of graphics, branding elements, labels, and any other visual aspects of the packaging. More complex designs or the use of specialized printing techniques can increase costs.

4. **Tooling and Equipment**: For certain types of packaging, custom tools, molds, or equipment might be needed. These initial setup costs can significantly impact the total packaging cost, but they can be spread over a larger production volume to reduce the cost per unit.

5. **Transportation and Logistics**: Packaging costs can also include expenses related to transporting the packaging materials to the production facility and the final packaged products to distribution centers or retailers.

6. **Quality Control and Testing**: Ensuring that the packaging meets quality standards may require additional testing and inspections, which can add to the overall cost.
7. **Sustainability Efforts**: Investing in sustainable and eco-friendly packaging materials may involve higher upfront costs, but businesses may find long-term cost savings through enhanced brand reputation and reduced environmental impacts.
8. **Storage and Warehousing**: If a large volume of packaging materials needs to be stored or warehoused before use, there will be associated costs.
9. **Regulatory Compliance**: Compliance with packaging regulations and standards set by governments or industry bodies may require additional testing or certifications, affecting overall costs.
10. **Waste Management and Recycling**: Proper waste management and recycling processes might be needed for certain packaging materials, which can influence costs.
11. **Customization and Special Features**: Packaging with unique features or functionalities, such as interactive elements or smart packaging technology, will typically incur higher costs.
12. **Packaging Volume**: Economies of scale can impact the overall packaging cost. Higher production volumes often lead to lower unit costs.

It's essential for businesses to carefully assess their packaging needs and consider various cost factors while striking a balance between functional requirements, aesthetics, and budget considerations. Sustainable packaging initiatives can also lead to cost savings over time, as consumers increasingly value environmentally friendly products and packaging.

1.7 Economy in Packaging

The economy in packaging refers to the efficient and cost-effective utilization of resources, materials, and processes in the packaging industry. It involves optimizing various aspects of packaging to minimize costs while maintaining or enhancing the functionality and appeal of the packaging. Here are some key elements of the economy in packaging:

1. **Material Selection**: Choosing appropriate packaging materials that meet the specific requirements of the product while considering cost implications is essential. Finding the right balance between cost, functionality, and sustainability is crucial in the economy of packaging.

2. **Design Efficiency**: An efficient packaging design maximizes the use of available space and reduces material waste. A well-thought-out design can minimize the need for excess cushioning or secondary packaging, thereby reducing costs.

3. **Standardization**: Standardizing packaging sizes and formats can lead to cost savings in production,

transportation, and storage. It allows for better optimization of resources and simplifies supply chain logistics.

4. **Economies of Scale**: Producing packaging materials in large quantities can lead to lower unit costs. This is especially true for materials like cardboard boxes, plastic bags, or labels.

5. **Automation and Technology**: Employing automation and advanced technology in packaging processes can increase efficiency, reduce labor costs, and minimize human errors.

6. **Sustainable Packaging**: Although sustainable packaging may have higher upfront costs, it can lead to long-term savings by improving brand reputation, reducing waste disposal costs, and attracting eco-conscious consumers.

7. **Reducing Excess Packaging**: Over-packaging a product can lead to unnecessary costs in materials, transportation, and waste management. Right-sizing and minimizing packaging where possible can result in cost savings.

8. **Recycling and Reuse**: Emphasizing recyclability and reusability of packaging materials can help reduce the need for new raw materials and lower waste management costs.

9. **Transportation Efficiency**: Optimizing packaging dimensions and weights can lead to more efficient

transportation, reducing fuel costs and environmental impacts.

10. **Supplier Relationships**: Maintaining strong relationships with packaging suppliers can lead to cost benefits through bulk purchasing, negotiated prices, and access to innovative packaging solutions.

11. **Regulatory Compliance**: Being aware of and complying with packaging regulations can prevent costly fines or penalties.

12. **Waste Reduction**: Implementing waste reduction strategies and lean manufacturing principles can minimize material wastage and associated costs.

The economy in packaging is a multifaceted approach that requires careful consideration of various factors, including materials, design, production processes, transportation, and end-of-life management. By focusing on efficiency, sustainability, and cost-effectiveness, businesses can achieve optimal packaging solutions that benefit both their bottom line and the environment.

1.8 Freight Vs Packaging

Freight and packaging are two essential aspects of the logistics and supply chain process, each serving distinct but interconnected purposes:

Packaging:

Packaging refers to the process of enclosing and protecting products with suitable materials to ensure safe transportation, storage, and delivery to the end consumer. Packaging plays a vital role in preserving product integrity, preventing damage during transit, and providing necessary information to consumers. Some key points about packaging include:

1. **Protection**: Packaging safeguards products from external factors like impact, moisture, temperature changes, and contamination.
2. **Branding and Marketing**: Packaging is a powerful tool for branding and marketing, as it creates a visual identity and influences consumer perception.
3. **Safety and Compliance**: Packaging must adhere to safety regulations and industry standards to ensure that products are safe for consumers to use.
4. **Materials and Design**: The choice of packaging materials and design depends on the product's nature, fragility, transportation requirements, and environmental impact.

Freight:

Freight refers to the goods or products being transported from one location to another, typically by various modes of transportation, such as trucks, ships, airplanes, or trains. Freight logistics involve planning, organizing, and managing the movement of goods throughout the supply chain. Key points about freight include:

1. **Transportation**: Freight involves the physical movement of goods from the point of origin to the destination.

2. **Efficiency and Cost**: Efficient freight logistics aim to optimize routes, reduce transit times, and minimize transportation costs.

3. **Handling and Storage**: Freight logistics also consider loading and unloading procedures, as well as proper storage during transit.

4. **Documentation and Compliance**: Freight logistics require accurate documentation, such as bills of lading and customs forms, to ensure compliance with legal and regulatory requirements.

5. **Tracking and Tracing**: Real-time tracking of freight helps monitor its movement, location, and estimated time of arrival.

The relationship between packaging and freight is interconnected. Well-designed and durable packaging is crucial for protecting products during transportation, reducing the risk of damage or losses. Effective packaging can contribute to efficient freight logistics by facilitating easier handling, stacking, and loading, ultimately saving time and transportation costs. Conversely, inefficient or inadequate packaging can lead to product damage, loss, or delays, negatively impacting freight logistics and increasing overall transportation expenses.

Ultimately, packaging and freight are essential components of the supply chain, and optimizing both aspects is critical for a successful and cost-effective logistics operation. By considering the interplay between packaging and freight, businesses can improve their overall efficiency, reduce waste, enhance customer satisfaction, and achieve better control over their supply chain processes.

1.9 Material Handling Vs Packaging And Weight Constraint

Material handling and packaging are closely related aspects of logistics and supply chain management, both playing crucial roles in ensuring the efficient movement and transportation of goods. Weight constraint is an important consideration that impacts both material handling and packaging decisions. Let's explore each of these concepts:

Material Handling:

Material handling refers to the process of moving, controlling, and managing goods and materials within a manufacturing facility, warehouse, or distribution center. It involves various activities such as loading, unloading, storing, picking, and transporting goods. The goal of material handling is to streamline operations, reduce costs, and enhance productivity. Key points about material handling include:

1. **Efficiency**: Effective material handling systems and equipment are designed to improve efficiency, reduce product damage, and optimize workflow.
2. **Safety**: Proper material handling practices prioritize the safety of workers and minimize the risk of accidents or injuries.
3. **Equipment and Technology**: Material handling equipment, such as forklifts, conveyors, and automated systems, are used to facilitate smooth material flow.
4. **Storage and Organization**: Material handling includes decisions on how products are stored, organized, and retrieved for efficient access and inventory management.
5. **Training and Skill**: Trained personnel are essential for effective material handling to ensure proper handling and use of equipment.

Packaging and Weight Constraint:

Packaging plays a critical role in protecting goods during transportation, minimizing damage, and ensuring that products reach their destination in good condition. Weight constraint is a consideration in both packaging design and material handling. Some points to consider:

1. **Product Protection**: Packaging must be able to withstand the weight of the product and any potential stacking or compression forces during transportation.

2. **Handling and Transportation**: The weight of the packaging itself impacts material handling, as heavier packaging may require specialized equipment or manual handling precautions.

3. **Shipping Costs**: Heavier packaging can increase transportation costs due to higher shipping weights and volume.

4. **Regulatory Compliance**: Some modes of transportation have weight restrictions and regulations that must be adhered to.

5. **Sustainability**: Reducing unnecessary packaging weight can contribute to sustainability efforts and minimize the overall environmental impact.

Finding the right balance between packaging protection and weight constraint is crucial. Over-packaging can lead to increased costs and environmental concerns, while under-packaging may result in product damage and losses during transit. It's essential to optimize packaging design to ensure it provides adequate protection while minimizing excessive weight.

By considering both material handling efficiency and weight constraint in packaging decisions, businesses can create a supply chain that operates smoothly, cost-effectively, and sustainably. Collaborating with experts in material handling and packaging design can help strike the right balance and improve the overall efficiency of the logistics and supply chain process.

1.10 Consumer Attracting Packaging

Consumer-attracting packaging, also known as attractive or eye-catching packaging, is designed to capture the attention of consumers and entice them to purchase the product. It is an important marketing tool that can significantly impact consumer behavior and influence buying decisions. Here are some key elements and strategies that contribute to consumer-attracting packaging:

1. **Visual Appeal**: Eye-catching packaging often features vibrant colors, bold graphics, and visually appealing designs. High-quality images and artwork can create an emotional connection with consumers.

2. **Unique Shape and Structure**: Packaging with distinctive shapes or innovative structures stands out on store shelves, drawing attention to the product.

3. **Branding Elements**: Clear and prominent branding elements, such as logos and brand colors, help consumers recognize and connect with the product and brand.

4. **Product Imagery**: Including images of the product itself or showcasing its usage can give consumers a better understanding of what they are buying.

5. **Clear Messaging**: Concise and compelling messaging communicates the product's key benefits or unique selling points, helping consumers make quick purchasing decisions.

6. **Eco-Friendly Packaging**: Packaging that highlights its eco-friendly features and sustainability appeals to consumers who prioritize environmental consciousness.

7. **Limited Editions and Seasonal Packaging**: Limited editions and seasonal packaging can create a sense of urgency and exclusivity, encouraging consumers to make a purchase.

8. **Transparency**: Transparent or windowed packaging allows consumers to see the product inside, enhancing product visibility and trust.

9. **Convenience and Functionality**: Packaging that offers ease of use, resealable features, or portable options can attract consumers seeking convenience.

10. **Texture and Finishes**: Incorporating unique textures, embossing, or special finishes (e.g., matte, gloss, foil) can create a tactile experience and make the packaging stand out.

11. **Storytelling**: Packaging that tells a compelling brand story or conveys the product's origin and journey can evoke emotional connections with consumers.

12. **Targeted Audience**: Tailoring packaging designs to specific target audiences helps appeal to the preferences and tastes of the intended consumers.

13. **Minimalism**: Simple and clean designs can also be attractive, especially to consumers looking for a modern and elegant aesthetic.

14. **Personalization**: Customizable or personalized packaging can make consumers feel special and valued.

Ultimately, consumer-attracting packaging should align with the brand identity, product positioning, and the preferences of the target market. By understanding the desires and motivations of their consumers, businesses can create packaging that not only captures attention but also drives engagement and fosters brand loyalty.

1.11 Packaging for safety

Packaging for safety is a critical aspect of product packaging that focuses on ensuring the protection of both the product and the consumers during storage, transportation, and handling. The primary goal of safety-oriented packaging is to minimize the risk of harm, accidents, and product damage. Here are some key considerations for packaging designed with safety in mind:

1. **Product Containment**: Safety-oriented packaging should securely contain the product to prevent spillage, leakage, or any other form of product escape that could be hazardous.

2. **Tamper-Evidence**: Packaging that provides clear evidence of tampering ensures that consumers can

identify if the product has been compromised, helping to protect against tampering and contamination.

3. **Child-Resistant Packaging**: Certain products, such as medicines or household chemicals, may require child-resistant packaging to reduce the risk of accidental ingestion or exposure.

4. **Protective Barriers**: Packaging should include protective barriers to shield the product from external factors like moisture, light, temperature changes, and contaminants.

5. **Cushioning and Shock Absorption**: For fragile or sensitive products, proper cushioning and shock-absorbing materials help protect them from damage during transit.

6. **Proper Labeling**: Safety-oriented packaging includes clear and accurate labeling with information such as cautionary statements, handling instructions, and warning symbols.

7. **Durability**: Packaging materials should be durable enough to withstand the rigors of transportation and handling without compromising product integrity.

8. **Compliance with Regulations**: Safety-oriented packaging must adhere to relevant safety regulations and industry standards, ensuring that it meets all necessary safety requirements.

9. **Stability and Stackability**: Packaging should be designed for stability during storage and transportation, allowing for safe stacking to prevent tipping or falling.

10. **Emergency Handling Information**: Packaging may include emergency handling information to guide consumers in case of accidental exposure or spillage.
11. **Resealable Features**: Certain products benefit from resealable packaging to maintain product freshness and prevent spills after the initial opening.
12. **Handling Instructions**: Clear instructions on proper handling, storage, and disposal of the product can prevent accidents and misuse.
13. **Safety Certifications**: Packaging materials with safety certifications or compliance with safety standards can provide reassurance to consumers.
14. **Environmentally Friendly**: Sustainable packaging options can contribute to safety by reducing harmful environmental impacts and supporting eco-friendly practices.

Safety-oriented packaging is especially crucial for products with potential hazards, such as pharmaceuticals, chemicals, sharp objects, and electronics. By prioritizing safety in packaging design and materials, businesses can build trust with consumers, protect their brand reputation, and ensure the well-being of both customers and employees throughout the supply chain.

1.12 Packaging to boost sale

Packaging plays a significant role in boosting sales by attracting consumers, creating brand recognition, and influencing purchasing decisions. A well-designed and strategically crafted packaging can make products stand out on store shelves and enhance their perceived value. Here are some packaging strategies that can help boost sales:

1. **Eye-Catching Design**: Packaging with visually appealing graphics, colors, and imagery can capture consumers' attention and create a positive first impression.

2. **Branding and Logo Visibility**: Prominent branding elements and a recognizable logo on the packaging help reinforce brand identity and build brand loyalty.

3. **Unique Packaging Shapes**: Innovative and distinctive packaging shapes can make products memorable and stand out from competitors.

4. **Clear Product Benefits**: Packaging should clearly communicate the product's benefits and unique selling points, making it easier for consumers to understand why they should choose it over others.

5. **Limited Editions and Seasonal Packaging**: Limited editions or special seasonal packaging can create a sense of urgency and exclusivity, encouraging consumers to make a purchase.

6. **Storytelling**: Packaging that tells a compelling brand story or showcases the product's journey can create an emotional connection with consumers.

7. **Product Imagery**: Including high-quality images of the product in use or showcasing its features can give consumers a better idea of what they are buying.

8. **Packaging for Different Consumer Segments**: Tailoring packaging to different consumer segments based on demographics, preferences, or usage occasions can resonate better with target audiences.

9. **Convenience and Functionality**: Packaging that is easy to open, use, and store can enhance the overall customer experience and drive repeat purchases.

10. **Eco-Friendly Packaging**: Environmentally friendly and sustainable packaging appeals to eco-conscious consumers, who may be more inclined to choose products with eco-friendly packaging.

11. **Visual Hierarchy**: Prioritize key information, such as product name, benefits, and branding, using visual hierarchy to ensure that essential details are easily noticed.

12. **Package Size and Quantity**: Offering packaging options that suit different consumer needs, such as family-sized or single-serving packs, can widen the product's appeal.

13. **In-Store Visibility**: Ensuring that the packaging design stands out on store shelves, with clear differentiation from competitors, is crucial for attracting impulse buyers.

14. **Cross-Promotions and Bundling**: Packaging can be used for cross-promotions or bundling of products, encouraging consumers to try complementary items or larger packs.

15. **Testimonials or Reviews**: Displaying positive customer testimonials or reviews on the packaging can build trust and credibility.

By strategically incorporating these elements into the packaging design, businesses can increase their products' visibility, customer engagement, and overall sales potential. It's important to keep consumer preferences, market trends, and brand values in mind while designing packaging that resonates with the target audience and drives successful sales outcomes.

1.13 Strengthening Methods

Strengthening methods in packaging are techniques and strategies aimed at enhancing the durability, protection, and overall quality of the packaging to ensure the safe transportation and preservation of products. These methods help minimize the risk of damage, breakage, or spoilage during transit and handling. Here are some common strengthening methods used in packaging:

1. **Reinforced Materials**: Using stronger and more durable materials, such as corrugated cardboard or thicker plastics, can improve the structural integrity of the packaging.

2. **Cushioning and Padding**: Incorporating cushioning materials, such as foam inserts, air-filled bubbles, or molded pulp, provides extra protection against impact and shock during transportation.

3. **Inner Packaging**: Separately packaging individual items within the main packaging can prevent collisions or friction that might lead to damage.

4. **Dividers and Partitions**: Adding dividers or partitions within the packaging can prevent products from shifting and bumping into each other.

5. **Corner and Edge Protectors**: Applying corner and edge protectors can reinforce vulnerable points of the packaging, reducing the risk of damage in those areas.

6. **Tamper-Proof Seals**: Including tamper-evident seals provides a visual indication of the product's integrity and ensures that the packaging remains intact until it reaches the end consumer.

7. **Overwrapping and Shrink Wrapping**: Applying an additional layer of protective overwrapping or shrink wrapping can enhance the resistance to moisture and tampering.

8. **Waterproofing**: Using waterproof coatings or materials can safeguard the product from damage caused by exposure to moisture or humidity.

9. **Vacuum Packaging**: Vacuum sealing removes air from the packaging, creating a tight seal that preserves the freshness and quality of perishable products.

10. **Protective Sleeves and Covers**: Adding protective sleeves or covers can shield products from external elements and environmental hazards.

11. **Barcoding and Tracking**: Implementing barcode labels and tracking systems allows for better visibility and control throughout the supply chain, reducing the risk of misplacement or mishandling.

12. **Testing and Quality Control**: Conducting tests, such as drop tests or compression tests, ensures that the packaging meets specified strength requirements.

13. **Custom Packaging**: Tailoring packaging to fit the product's shape and size precisely reduces movement and provides a snug fit, minimizing the risk of damage.

14. **Use of Pallets and Strapping**: Palletizing products and securing them with strapping or stretch wrap can increase stability and protection during transportation.

Strengthening methods in packaging should be chosen based on the specific needs and characteristics of the product being transported. By investing in robust and protective packaging, businesses can reduce product losses, enhance customer satisfaction, and maintain brand reputation by delivering products in optimal condition.

1.14 Packaging Functions

Packaging serves multiple essential functions throughout the product lifecycle, from manufacturing to consumer use. These functions are crucial for protecting the product, facilitating transportation, and enhancing the overall customer experience. Here are the primary functions of packaging:

1. **Protection**: One of the fundamental functions of packaging is to protect the product from damage, contamination, and external elements such as moisture, dust, and light. Packaging acts as a barrier, ensuring that the product reaches the consumer in optimal condition.
2. **Containment**: Packaging securely encloses and contains the product, preventing spillage, leakage, or escape during handling, transportation, and storage.
3. **Preservation**: Packaging can extend the shelf life of perishable products by providing a controlled environment that preserves freshness and prevents spoilage.
4. **Identification and Information**: Packaging includes labels, barcodes, and product information that help

identify the product, its contents, usage instructions, and safety precautions.

5. **Convenience and Handling**: Packaging is designed for ease of handling, storage, and transportation. It may include features like handles, resealable closures, or tear strips for consumer convenience.
6. **Marketing and Branding**: Packaging is a powerful marketing tool that conveys the brand's identity, values, and messaging. It attracts attention, differentiates the product from competitors, and influences consumer purchasing decisions.
7. **Product Differentiation**: Unique and innovative packaging designs help distinguish products in a crowded market, making them more memorable to consumers.
8. **Security and Tamper Resistance**: Packaging can include security features, such as tamper-evident seals or anti-counterfeiting measures, to ensure product authenticity and consumer safety.
9. **Environmental Sustainability**: Eco-friendly packaging solutions contribute to reducing the environmental impact by promoting recycling, using biodegradable materials, or reducing packaging waste.
10. **Information Communication**: Packaging communicates important product details, such as nutritional information, ingredients, product warnings, and usage guidelines.
11. **Transportation and Distribution**: Packaging is designed to facilitate transportation and

distribution processes, ensuring safe and efficient handling throughout the supply chain.

12. **Storage and Inventory Management**: Packaging allows for organized storage and inventory management, facilitating stock replenishment and supply chain efficiency.
13. **User Experience**: Packaging can enhance the overall customer experience, influencing consumer satisfaction and loyalty to the brand.
14. **Regulatory Compliance**: Packaging must adhere to relevant regulations and standards, ensuring that it meets safety and legal requirements.
15. **Customization**: Packaging can be customized to cater to specific product needs, target audiences, and market preferences.

By fulfilling these diverse functions, packaging plays a critical role in product protection, consumer engagement, and the success of businesses in the global market. Effective packaging design considers all these functions while balancing aesthetics, functionality, and environmental impact.

1.15 Benefits And Challenges

Packaging offers various benefits to both businesses and consumers, but it also presents certain challenges. Let's explore the advantages and challenges of packaging:

Benefits of Packaging:

1. **Product Protection**: Packaging safeguards products from damage, contamination, and spoilage during transportation, storage, and handling.

2. **Brand Recognition and Marketing**: Attractive and well-designed packaging can increase brand visibility, differentiate products from competitors, and influence purchasing decisions.

3. **Information and Communication**: Packaging provides essential product information, usage instructions, safety warnings, and nutritional details, helping consumers make informed choices.

4. **Convenience and Functionality**: Packaging can be designed for easy handling, portability, resealing, and portion control, enhancing the overall consumer experience.

5. **Preservation and Shelf Life Extension**: Packaging materials with barrier properties can extend the shelf life of perishable products by controlling environmental factors like moisture, light, and oxygen.

6. **Product Differentiation**: Unique packaging designs and innovative shapes can make products stand out on store shelves, attracting consumer attention.

7. **Consumer Safety**: Packaging can incorporate tamper-evident features and child-resistant closures to enhance consumer safety and prevent accidents.

8. **Supply Chain Efficiency**: Packaging facilitates organized storage, transportation, and inventory management, streamlining supply chain processes.

9. **Sustainability Initiatives**: Eco-friendly packaging options contribute to reducing environmental impact and appeal to environmentally conscious consumers.

Challenges of Packaging:

1. **Environmental Impact**: Packaging waste contributes to environmental pollution, and non-recyclable or single-use plastics can lead to long-lasting ecological harm.

2. **Excessive Packaging**: Over-packaging can lead to unnecessary waste, higher production costs, and increased transportation expenses.

3. **Packaging Costs**: High-quality and specialized packaging materials can add to product costs, impacting profit margins.

4. **Regulatory Compliance**: Meeting packaging regulations and standards can be complex and time-consuming, requiring additional testing and certifications.

5. **Packaging Waste Management**: Proper disposal and recycling of packaging waste can be challenging for consumers and municipalities.

6. **Packaging Design Complexity**: Intricate packaging designs may require specialized equipment and processes, increasing production complexity.
7. **Consumer Perception**: Poorly designed or unsustainable packaging can lead to negative consumer perceptions and affect brand reputation.
8. **Shelf Space and Retailer Demands**: Limited shelf space and specific retailer requirements can influence packaging decisions and add constraints for manufacturers.
9. **Innovation and Trends**: Keeping up with packaging trends and evolving consumer preferences can be challenging for businesses.

Balancing the benefits and challenges of packaging is a crucial task for businesses seeking to deliver quality products while being environmentally responsible. Sustainable packaging initiatives, waste reduction strategies, and continuous innovation in design and materials can help address some of the challenges and create a positive impact on both the business and the environment.

OBJECTIVE QUESTION

1. What is the primary need for packaging?

a) Protection b) Transportation

c) Branding d) Storage

Answer: a) Protection

2. Which of the following is NOT a nature of packaging?

a) Convenience b) Marketing

c) Aesthetic appeal d) Natural origin

Answer: d) Natural origin

3. Which material is commonly used for packaging beverages like soft drinks and water?

a) Glass b) Plastic

c) Paper d) Metal

Answer: b) Plastic

4. Which type of packaging is designed to be used multiple times?

a) Single-use packaging b) Disposable packaging

c) Reusable packaging d) Biodegradable packaging

Answer: c) Reusable packaging

5. Which component of cost is NOT typically involved in packaging?

a) Labor b) Raw materials

c) Marketing d) Transportation

Answer: c) Marketing

6. Economy in packaging aims to strike a balance between:

a) Cost and aesthetics b) Cost and functionality

c) Cost and branding d) Cost and convenience

Answer: b) Cost and functionality

7. When comparing freight and packaging, which one directly affects transportation costs?

a) Freight b) Packaging

c) Both equally d) Neither

Answer: b) Packaging

8. What does "material handling" primarily refer to in the context of packaging?

a) Designing materials b) Storing materials

c) Moving materials within a facility d) Packaging materials

Answer: c) Moving materials within a facility

9. Which constraint is often considered when designing packaging for products?

a) Color b) Size

c) Shape d) Weight

Answer: d) Weight

10. Consumer-attracting packaging aims to:

a) Reduce product visibility

b) Minimize brand recognition

c) Capture consumer attention

d) Decrease product shelf life

Answer: c) Capture consumer attention

11. Packaging primarily serves the purpose of ensuring:

a) Product freshness b) Product profitability

c) Product distribution d) Product development

Answer: a) Product freshness

12. What type of packaging is designed to encourage impulse buying and increase sales?

a) Protective packaging

b) Sales-oriented packaging

c) Eco-friendly packaging

d) Minimalist packaging

Answer: b) Sales-oriented packaging

13. Strengthening methods in packaging are used to:

a) Make the packaging more attractive

b) Enhance the shelf life of the product

c) Protect the product during transportation and handling

d) Reduce packaging costs

Answer: c) Protect the product during transportation and handling

14. What is one of the key functions of packaging?

a) Reducing production costs

b) Enhancing environmental sustainability

c) Communicating product information

d) Increasing raw material usage

Answer: c) Communicating product information

15. Which type of packaging is used primarily for delicate and fragile items?

a) Bulk packaging

b) Primary packaging

c) Secondary packaging

d) Tertiary packaging

Answer: b) Primary packaging

Word Limit (80-100)

1. **What is the packaging ? explain the nature of packaging?**
2. **What are the primary types of packaging methods and their applications in different sectors?**
3. **What are the key components of cost involved in packaging production and distribution?**
4. **How does the concept of economy in packaging relate to the trade-off between freight costs and packaging expenses?**
5. **What is advantage and challenges of packaging?**
6. **How can reusable materials be effectively incorporated into packaging designs?**

Word Limit (180-200)

1. **Can you provide insights into the diverse materials used in packaging and their suitability for various types of products, considering environmental sustainability and product protection?**
2. **How can reusable materials be effectively integrated into packaging design, and what are the key advantages and challenges associated with their adoption in the packaging industry?**
3. **In what ways does the type of packaging impact the cost involved in product distribution, and how can businesses strike a balance between cost-efficiency and product protection through their packaging choices?**

UNIT-2. Special Packaging

LESSON OUTLINE-

- Packaging For Special Product
- Packaging For Cold Product, Freezing Product,
- Dairy Product And Packaging For Hot Products
- Fruits /Vegetable Packaging,
- Packaging Of Grains,
- Packaging Of Pharmacy Products And
- Food Products
- Packaging Of Liquids, Water, Edible Oil,
- Packaging Of Cosmetics And Powdered Articles
- Packaging Guide Lines,
- Single Box And Box In Box Packaging –Storing In Cold / Frozen Storage.

2.1 Packaging for special product

Packaging for special products requires careful consideration and customization to meet the unique requirements of the specific product. Special products can include high-value items, fragile or perishable goods, hazardous materials, luxury items, or unique promotional products. Here are some key factors to consider when designing packaging for special products:

1. **Product Protection**: The primary function of packaging is to protect the special product from

damage, breakage, or contamination during transportation and handling. Choose packaging materials and cushioning that offer the necessary level of protection.

2. **Customization**: Tailor the packaging design to fit the dimensions and shape of the special product precisely. Custom packaging ensures a snug fit, minimizing movement and reducing the risk of damage.

3. **Handling and Transportation**: Consider the logistics of handling and transportation when designing packaging. For fragile or delicate items, use materials that can withstand impacts and vibrations during transit.

4. **Temperature and Environmental Factors**: If the special product is sensitive to temperature or humidity, use packaging materials with appropriate barrier properties to maintain product integrity.

5. **Security and Tamper Resistance**: For high-value or sensitive products, incorporate security features like tamper-evident seals or RFID tags to ensure product authenticity and prevent tampering.

6. **Branding and Presentation**: Packaging for special products is an opportunity to reinforce branding and create a memorable unboxing experience for customers. Consider using high-quality materials, unique finishes, and custom printing.

7. **Compliance and Regulations**: Ensure that the packaging meets all relevant regulatory requirements, especially if dealing with hazardous or restricted materials.

8. **Sustainability**: Where possible, opt for sustainable packaging solutions that align with the environmental values of your brand and customers.

9. **User Experience**: Consider the convenience and usability of the packaging from the customer's perspective. Packaging should be easy to open and reseal if applicable.

10. **Testing and Quality Assurance**: Conduct thorough testing and quality assurance checks to ensure that the packaging can withstand the challenges of transportation and handling.

11. **Limited Editions and Collectibles**: For special limited edition or collectible products, design packaging that reflects the exclusivity and uniqueness of the item, creating a sense of value for collectors.

12. **Promotional Packaging**: If the special product is part of a promotional campaign or gift set, design packaging that complements the theme and enhances the overall presentation.

13. **Localization**: Consider cultural preferences and local regulations when designing packaging for special products intended for international markets.

Packaging for special products is an opportunity to elevate the brand, enhance the customer experience, and provide added value to the consumer. By carefully considering the unique characteristics of the product and the target audience, businesses can create packaging that effectively showcases the product and aligns with the brand's identity and values.

2.2 Packaging For Cold Product, Freezing Product, Dairy Product

Packaging for cold, freezing, and dairy products requires specific considerations to maintain product quality, freshness, and safety throughout the supply chain. These products are highly sensitive to temperature and environmental conditions, making appropriate packaging crucial. Here are some key factors to consider for each type of product:

Packaging for Cold Products (Chilled Products):

1. **Insulation**: Use insulated packaging materials to maintain the desired temperature and prevent temperature fluctuations during transportation.

2. **Temperature Indicators**: Incorporate temperature indicators to monitor and ensure that the cold chain is maintained throughout the distribution process.

3. **Gel Packs or Ice Packs**: Include gel packs or ice packs to keep the product cool during transit and storage.

4. **Airflow**: Design packaging to allow for proper airflow, ensuring even distribution of cold air and preventing condensation.

5. **Water-Resistant**: Ensure that the packaging is water-resistant to protect the product from moisture and potential leaks.

6. **Custom Fit**: Customized packaging that snugly fits the product helps maintain temperature control and prevents shifting during transportation.

7. **Quick Access**: Design packaging for easy access to the product without compromising its temperature integrity.

Packaging for Freezing Products (Frozen Products):

1. **Freezer-Safe Materials**: Use packaging materials that can withstand freezing temperatures without becoming brittle or compromised.

2. **Moisture Barrier**: Provide a strong moisture barrier to prevent freezer burn and maintain the product's texture and taste.

3. **Air-Tight Sealing**: Ensure the packaging is air-tight to prevent ice crystal formation and maintain product quality.

4. **Stackable Design**: Consider stackable packaging to optimize freezer space during storage and transportation.

5. **Thawing Instructions**: Include clear instructions on how to thaw the product safely without compromising its quality.

6. **Package Strength**: Ensure the packaging is durable enough to withstand freezing conditions and handling.

Packaging for Dairy Products:

1. **Temperature Control**: Dairy products are temperature-sensitive, so packaging must provide proper insulation to maintain their freshness.

2. **Light Protection**: Dairy products are sensitive to light exposure, so use opaque or tinted packaging to prevent light-induced spoilage.

3. **Leak-Proof**: Dairy packaging should be leak-proof to avoid spills and contamination.

4. **Easy Dispensing**: Consider packaging that allows for easy pouring or dispensing, especially for liquids like milk or yogurt.

5. **Portion Control**: For individual servings, consider packaging with portion control features to reduce waste and ensure consistent servings.

6. **Reclosable**: Use reclosable packaging to keep the product fresh after opening.

7. **Hygiene and Safety**: Packaging should meet hygiene standards to ensure the safety and quality of dairy products.

8. **Shelf Life Information**: Include clear information about the product's expiration date and storage conditions.

9. **Environmentally Friendly**: Opt for eco-friendly packaging options to align with sustainability goals.

When designing packaging for these sensitive products, businesses must also comply with relevant food safety regulations and industry standards to ensure consumer safety. Properly designed packaging helps preserve product quality, reduce food waste, and enhance the overall customer experience.

2.3 Packaging for hot products

Packaging for hot products requires careful consideration to ensure safety, retain product quality, and protect consumers from potential burns or accidents. Hot products can include food items like ready-to-eat meals, hot beverages, or heat-generating items like electronic devices or heating pads. Here are some key factors to consider when designing packaging for hot products:

1. **Heat Resistance**: Use materials that can withstand high temperatures without deforming or releasing harmful chemicals.

2. **Insulation**: Design packaging with insulating properties to retain heat and keep the product at the desired temperature.

3. **Heat Retention**: Ensure that the packaging can retain heat for an extended period to maintain product freshness and quality.
4. **Heat Ventilation**: For products that emit heat, provide ventilation to dissipate excess heat and prevent overheating of the packaging.
5. **Safety Features**: Incorporate safety features, such as heat-resistant handles or grips, to protect consumers from burns when handling the hot product.
6. **Sealing Integrity**: Ensure that the packaging is well-sealed to prevent leaks or spills, especially for liquid hot products.
7. **Barrier Properties**: Use materials with good barrier properties to prevent heat transfer to the external environment and maintain the product's temperature.
8. **Tamper-Evidence**: Include tamper-evident features to assure consumers that the product has not been tampered with and is safe to consume.
9. **Heat-Resistant Labels**: Use heat-resistant labels and inks to prevent label damage or fading due to high temperatures.
10. **Thermal Performance Testing**: Conduct thermal performance testing to validate the packaging's ability to retain heat effectively.
11. **User-Friendly Design**: Design packaging with user-friendly features, such as easy-open

mechanisms, pour spouts, or tear strips for convenient access to the hot product.

12. **Custom Fit**: Customized packaging that securely holds the product in place reduces movement and minimizes the risk of spills or accidents.

13. **Regulatory Compliance**: Ensure that the packaging complies with safety regulations and guidelines for products intended to be consumed hot.

14. **Portion Control**: For single-serve hot products, consider portion control features to avoid waste and ensure consistent servings.

15. **Insulated Outer Packaging**: For products with extremely high temperatures, use an insulated outer packaging layer to provide an extra layer of protection for consumers.

Safety is of utmost importance when designing packaging for hot products. Manufacturers should conduct thorough testing to ensure that the packaging effectively retains heat and meets safety standards. Properly designed packaging not only protects consumers but also enhances the overall customer experience and product quality.

2.4 Fruits /vegetable Packaging

Packaging for fruits and vegetables is essential to ensure their freshness, quality, and safety throughout the supply chain. It helps protect the produce from physical damage, contamination, and spoilage while also enhancing its visual appeal for consumers. Here are some common packaging methods used for fruits and vegetables:

1. **Produce Bags**: Thin, lightweight, and breathable plastic or mesh bags are commonly used for packaging loose fruits and vegetables. These bags allow air circulation, preventing condensation and reducing the risk of mold growth.

2. **Clamshell Containers**: These hinged plastic containers are used to package delicate fruits like berries, grapes, or cherry tomatoes. They provide protection and allow for easy display at the retail level.

3. **Punnets**: Punnets are small, shallow plastic trays used for packaging fruits like strawberries, blueberries, or cherry tomatoes. They are often covered with a clear plastic lid to protect the produce and allow visibility.

4. **Net Bags**: These are mesh bags used for packaging larger fruits like oranges, onions, or potatoes. Net bags provide good ventilation and allow easy handling of the produce.

5. **Trays with Film Wrapping**: Some fruits and vegetables are packaged on trays and covered with a plastic film to offer protection and maintain freshness.

6. **Modified Atmosphere Packaging (MAP)**: This method involves adjusting the gas composition inside

the packaging to extend the shelf life of the produce. It is commonly used for products like pre-cut fruits and vegetables.

7. **Bulk Bins**: Large bins or crates are used for displaying and selling loose produce in larger quantities. Customers can fill smaller bags or containers themselves.
8. **Corrugated Cardboard Boxes**: Sturdy cardboard boxes are used for transporting and storing bulk quantities of fruits and vegetables. These boxes protect the produce during transportation.
9. **Reusable Produce Bags**: Environmentally friendly, reusable bags made of cloth or mesh are becoming more popular as an alternative to single-use plastic bags.
10. **Pre-Packaged Salad Bags**: For pre-packaged salads, a combination of leafy greens and other vegetables is placed in a bag with resealable features for convenience.
11. **Labels and Stickers**: Packaging often includes labels with product information, country of origin, barcodes, and pricing details.

Packaging for fruits and vegetables must strike a balance between protecting the produce and minimizing environmental impact. Sustainable and biodegradable packaging options are gaining popularity as consumers and retailers seek eco-friendly alternatives to traditional plastic packaging. Additionally, packaging designs that allow for proper ventilation and easy handling help preserve the freshness and quality of the produce, ultimately contributing to customer satisfaction and reduced food waste.

2.5 Packaging of grains

Packaging of grains is essential to protect them from moisture, pests, and contamination, ensuring their quality and safety throughout storage, transportation, and distribution. Properly packaged grains also facilitate easy handling and measurement. Here are some common packaging methods used for grains:

1. **Bags**: Grain bags are a common packaging option, typically made of woven polypropylene or jute. These bags offer protection from moisture, pests, and physical damage. They come in various sizes to accommodate different quantities of grains.

2. **Bulk Containers**: For large-scale storage and transportation, bulk containers like flexible intermediate bulk containers (FIBCs) or bulk bins are used. These containers can hold significant quantities of grains efficiently.

3. **Lined Cartons**: Cartons lined with moisture-resistant materials, such as plastic liners or aluminum foil, are used for packaging grains like rice or cereal.

4. **Plastic Containers**: Durable, airtight plastic containers are suitable for storing smaller quantities of grains, keeping them fresh and protected.
5. **Vacuum-Sealed Bags**: Vacuum sealing removes air from the packaging, creating a tight seal that helps preserve the quality and freshness of grains.
6. **Hermetically Sealed Bags**: These bags are air-tight and provide an extra layer of protection against pests and moisture.
7. **Pouches with Zipper Closures**: Resealable pouches with zipper closures are convenient for packaging smaller quantities of grains and allowing for easy access.
8. **Labels and Barcodes**: Packaging often includes labels with product information, batch numbers, expiration dates, and barcodes for inventory management and traceability.
9. **Sustainable Packaging**: Eco-friendly packaging options, such as compostable or biodegradable bags, are gaining popularity to reduce environmental impact.
10. **Grain Silos**: For bulk storage, grains are sometimes stored in silos, which are large containers designed to maintain the quality and condition of the grains over an extended period.

It's important to ensure that the packaging materials used for grains are food-grade, safe, and comply with food safety regulations. Additionally, proper handling and storage conditions are essential to maintain the grains' quality and prevent spoilage. For bulk storage, fumigation or temperature control measures may be necessary to protect against pests and fungi.

Well-designed packaging not only protects grains but also provides useful information to consumers and facilitates easy identification and handling. Effective grain packaging plays a crucial role in preserving food quality, reducing food waste, and ensuring a reliable food supply chain.

2.6 Packaging of pharmacy products

Packaging of pharmacy products, especially medications and healthcare items, requires special consideration to ensure their safety, efficacy, and compliance with regulatory requirements. Pharmacy product packaging serves several critical functions to protect both the product and the end consumer. Here are some key aspects of packaging for pharmacy products:

1. **Tamper-Evident Seals**: Pharmacy products should have tamper-evident seals to indicate whether the package has been opened or compromised, ensuring consumer safety.
2. **Child-Resistant Closures**: Medications and certain healthcare products may require child-resistant closures to prevent accidental ingestion by children.

3. **Light Protection**: Light-sensitive medications or products should be packaged in opaque containers or with light-resistant materials to preserve their stability.

4. **Moisture Barrier**: Medications sensitive to moisture should be packaged with materials that provide an effective moisture barrier to maintain their potency.

5. **Barrier to Oxygen**: Packaging should protect oxygen-sensitive pharmaceuticals from degradation by using barrier materials.

6. **Dosing and Usage Information**: Packaging must include clear dosing and usage instructions, warnings, and contraindications to ensure safe and appropriate use.

7. **Expiration Date**: Each package should display the product's expiration date to prevent the use of outdated medications.

8. **Unit-Dose Packaging**: Some medications are individually packaged in unit-dose formats to promote accurate dosing and reduce the risk of errors.

9. **Blister Packs**: Blister packs are commonly used to package individual doses of medications for easy administration and compliance monitoring.

10. **Prescription Labeling**: Prescription medications require specific labeling, including the

patient's name, prescribing physician, dosage instructions, and other relevant information.

11. **Serialization and Track-and-Trace**: To combat counterfeiting and ensure product traceability, serialization codes may be included on packaging.

12. **Temperature Control**: Temperature-sensitive pharmaceuticals may require insulated packaging or temperature-monitoring devices for transport.

13. **Sustainability**: Increasingly, eco-friendly and sustainable packaging options are being sought to minimize environmental impact.

14. **Adherence Aids**: Packaging may include adherence aids such as medication calendars or reminder systems to promote patient compliance.

15. **Multi-Product Packaging**: Combining multiple medications into a single package (e.g., blister strips with several medications) can enhance patient convenience.

Pharmacy product packaging must comply with stringent regulations, such as those set by the Food and Drug Administration (FDA) or other health regulatory authorities. Manufacturers and pharmacists must ensure that the packaging maintains product integrity, accurately communicates essential information, and provides adequate protection throughout the product's shelf life. Well-designed packaging is crucial to maintaining patient safety, ensuring proper medication use, and preventing medication errors.

2.7 Packaging of food products

Packaging of food products is critical to maintaining product freshness, ensuring food safety, providing essential information to consumers, and enhancing the overall consumer experience. Food packaging serves various functions and is tailored to the specific needs of different food items. Here are some common packaging types and considerations for food products:

1. **Primary Packaging**: This is the packaging that directly comes into contact with the food product. Examples include:
    - Bottles and jars for beverages, sauces, and condiments.
    - Pouches and sachets for snacks, powders, and liquids.
    - Trays and containers for ready-to-eat meals, salads, and deli items.
    - Cans for canned fruits, vegetables, and soups.
    - Clamshell containers for berries, cherry tomatoes, and other delicate produce.
2. **Secondary Packaging**: Secondary packaging is used to group and protect individual units of primary

packaging. It may also serve as a display unit in retail settings. Examples include:

- Cardboard boxes for holding multiple units of primary packaging (e.g., cereal boxes).
- Shrink wrap or stretch wrap for bundling cans or bottles together.

3. **Modified Atmosphere Packaging (MAP)**: This method involves altering the gas composition inside the packaging to extend the shelf life of the food product. It is commonly used for fresh produce, meats, and pre-packaged meals.
4. **Resealable Packaging**: Many food products, such as snacks and frozen items, come in resealable packaging to maintain freshness and allow for easy storage.
5. **Microwavable Packaging**: Some food products are packaged in materials that can be safely heated in a microwave, providing convenience to consumers.
6. **Tamper-Evident Seals**: To ensure food safety, many food products have tamper-evident seals or bands to indicate if the package has been opened or compromised.
7. **Informational Labels**: Food packaging includes labels with nutrition facts, ingredients, allergen information, storage instructions, expiration dates, and batch codes.

8. **Portion Control Packaging**: Single-serve packaging allows for portion control and reduces food waste.
9. **Sustainable and Eco-Friendly Packaging**: Environmentally friendly packaging options, such as compostable materials or recyclable packaging, are becoming increasingly popular.
10. **Easy-Open Features**: Packaging may include easy-to-open features like tear notches or pull tabs for consumer convenience.

Food packaging must comply with food safety regulations and industry standards to ensure that the product remains safe for consumption. It also plays a crucial role in branding and marketing, helping to differentiate products on store shelves and communicate key product information to consumers. Additionally, packaging helps protect food products during transportation and storage, reducing food spoilage and waste.

2.8 Packaging Of Liquids, Water, Edible Oil

Packaging liquids, such as water and edible oil, requires specialized packaging materials and design to prevent leakage, maintain product freshness, and ensure consumer safety. Here are some common packaging types and considerations for liquids:

Water Packaging:

1. **Bottles**: Water is often packaged in plastic bottles, typically made of PET (polyethylene terephthalate) or HDPE (high-density polyethylene). These bottles are

lightweight, durable, and come in various sizes, from small single-serve bottles to larger ones for home use.

2. **Closures**: The bottle caps or closures should be securely sealed to prevent leakage and maintain water purity. Many bottles use screw-on caps or flip-top caps for easy opening and resealing.

3. **Labeling**: Labels on water bottles provide essential information such as the brand, product name, water source, nutritional details (if applicable), and recycling information.

4. **Purity and Safety**: Water packaging should comply with safety and purity standards to ensure that the water is safe for consumption.

5. **Eco-Friendly Packaging**: Sustainable options, such as biodegradable or recyclable bottles, are becoming more popular to reduce environmental impact.

Edible Oil Packaging:

1. **Bottles or Containers**: Edible oils are typically packaged in glass or plastic bottles and containers. Glass bottles help preserve the oil's quality by protecting it from light exposure.

2. **Dark-Colored Bottles**: Some edible oils, like olive oil, are packaged in dark-colored glass bottles to protect the oil from light-induced oxidation.

3. **Tamper-Proof Seals**: Edible oil bottles often have tamper-proof seals to ensure product integrity and safety.
4. **Pour Spouts**: Some bottles include pour spouts or drip-free caps for controlled pouring and reducing mess.
5. **Labels and Certifications**: Labels provide information on the type of oil, production methods, nutritional facts, and any certifications (e.g., organic, non-GMO).
6. **Handle and Grip Design**: For larger containers, handles or grip features may be included to facilitate lifting and pouring.
7. **Bulk Packaging**: Edible oils are also packaged in larger containers, such as drums or IBC (Intermediate Bulk Container) tanks, for commercial and industrial use.

Packaging for liquids should be designed to prevent leakage during transportation and handling. Proper seals, closures, and cap designs are critical to maintaining product integrity. For consumer convenience, easy-to-open and resealable features are beneficial, especially for products that may be used over time.

Additionally, food-grade materials and compliance with food safety regulations are essential to ensure that the packaging does not contaminate the liquids and that the product remains safe for consumption. Sustainable packaging options, such as recyclable or biodegradable materials, are increasingly favored by environmentally conscious consumers and businesses.

2.9 Packaging Of Cosmetics And Powdered Articles

Packaging for cosmetics and powdered articles serves several purposes, including preserving product integrity, ensuring hygiene, providing product information, and enhancing the overall user experience. Here are some common packaging types and considerations for cosmetics and powdered articles:

Cosmetics Packaging:

1. **Containers and Jars**: Cosmetics such as creams, lotions, and lip balms are often packaged in containers or jars made of plastic, glass, or metal. These containers should be airtight and secure to prevent contamination and maintain product freshness.

2. **Pump Bottles**: Liquid foundations, serums, and lotions often come in pump bottles, which allow for controlled dispensing and minimize product waste.

3. **Tube Packaging**: Tubes are commonly used for products like lip gloss, mascara, and hand creams. They offer convenient application and precise dispensing.

4. **Compacts and Pans**: For pressed powders, blushes, and eyeshadows, compacts and pans with mirrors are popular packaging options for on-the-go use.

5. **Applicators and Brushes**: Some cosmetics packaging includes built-in applicators or brushes for easy and precise product application.

6. **Squeeze Tubes**: These are used for products with thicker consistencies, such as toothpaste or some types of skincare creams.

7. **Dispensing Caps**: Some bottles have dispensing caps that allow controlled pouring or dripping of liquid products.

8. **Labels and Branding**: Labels on cosmetics packaging provide information on the product name, ingredients, usage instructions, and branding.

9. **Tamper-Evident Seals**: Many cosmetics products have tamper-evident seals to assure consumers that the product has not been tampered with.

Powdered Articles Packaging:

1. **Jars and Containers**: Loose powder products like setting powders, blushes, and highlighters are often packaged in jars or containers with sifters to control product dispersion.

2. **Puff or Brush Applicators**: Some powdered articles come with attached puffs or brushes for easy and even application.

3. **Refillable Packaging**: Refillable compacts or palettes allow customers to replace the product once it's

finished, reducing waste and supporting sustainability.

4. **Tight Seals**: Powdered articles should have tight seals to prevent spills and leakage during transportation and use.

5. **Stackable Design**: For products sold as a set or collection, stackable packaging helps keep the items organized and compact.

6. **Travel-Friendly**: Packaging that is designed for travel, such as mini sizes or products with secure closures, is convenient for on-the-go use.

7. **Resealable Pouches**: Some powdered products are packaged in resealable pouches for easy storage and access.

Cosmetics and powdered articles packaging should be visually appealing and align with the brand's identity. Hygienic and easy-to-use packaging can improve the customer's overall satisfaction and experience with the product. Additionally, environmentally friendly packaging options, such as recyclable materials or refillable containers, are becoming more popular as consumers increasingly prioritize sustainability in their purchasing decisions.

2.10 Packaging Guide Lines

Packaging guidelines are essential to ensure that products are packaged safely, efficiently, and in compliance with relevant regulations. These guidelines may vary depending on the type of product, industry, and regional regulations. Here are some general packaging guidelines to consider:

1. **Product Protection**: Packaging should provide adequate protection to the product from physical damage, contamination, moisture, and other environmental factors during storage, transportation, and handling.

2. **Safety and Compliance**: Ensure that the packaging materials and design comply with all relevant safety standards, industry regulations, and legal requirements.

3. **Labeling and Information**: Include clear and accurate labels on the packaging with essential product information, such as product name, ingredients, usage instructions, allergen warnings, expiration date, and batch/lot numbers.

4. **Sustainability**: Whenever possible, opt for environmentally friendly packaging materials and practices to minimize the environmental impact and appeal to eco-conscious consumers.

5. **Cost-Efficiency**: Balance packaging quality with cost considerations to avoid unnecessary expenses while maintaining product integrity.

6. **Ease of Use**: Design packaging for user convenience, making it easy to open, handle, and reseal if applicable.

7. **Branding and Marketing**: Packaging should align with the brand's identity, creating a visually appealing and consistent brand image.
8. **Tamper Resistance**: Implement tamper-evident features to ensure the product's safety and authenticity.
9. **Transportation and Logistics**: Consider the practical aspects of packaging, such as stacking, palletization, and compatibility with transportation methods, to facilitate efficient shipping and distribution.
10. **Customization**: Tailor packaging solutions to fit the specific needs of the product, taking into account its size, shape, and characteristics.
11. **Quality Control**: Implement quality control measures to ensure that the packaging meets specified standards and remains consistent throughout production.
12. **Material Selection**: Choose packaging materials appropriate for the product's nature, ensuring compatibility and preventing interactions that might affect the product's quality.
13. **Product Presentation**: Packaging should enhance the product's visual appeal and provide a positive unboxing experience for customers.

14. **Accessibility and Inclusivity**: Consider packaging designs that are accessible and user-friendly for people with disabilities.

15. **Testing and Validation**: Conduct testing and validation procedures to assess the packaging's performance under different conditions, such as drop tests, compression tests, and environmental tests.

16. **Adherence to Industry Best Practices**: Stay informed about industry trends and best practices in packaging design, materials, and sustainability initiatives.

It's essential to review and update packaging guidelines regularly to adapt to changing consumer preferences, advances in materials technology, and evolving regulatory requirements. Collaborating with packaging experts and conducting consumer research can also provide valuable insights to optimize packaging solutions and create a positive impact on product success and consumer satisfaction.

2.11 Single box and box in box Packaging –storing in cold / frozen storage

Single box and box-in-box packaging are two different approaches to packaging products, and both can be used for storing items in cold or frozen storage.

Single Box Packaging:

Single box packaging refers to using a single layer of packaging to enclose the product. This can be a single carton or box that directly contains the product. For cold or frozen storage, single box packaging may include insulation or cold packs to maintain the desired temperature within the box. The packaging material should be selected to withstand the low temperatures without becoming brittle or compromised.

For example, if you have perishable food items that need to be stored in a cold environment, you may use a single box packaging with insulation material inside to keep the products chilled during transportation and storage.

Box-in-Box Packaging:

Box-in-box packaging involves placing the primary product box or container inside an outer box. This method provides an additional layer of protection and insulation, which can be particularly useful for storing items in cold or frozen environments.

For instance, in the case of frozen food products, the primary product packaging (such as a bag of frozen vegetables) would be placed inside an outer corrugated cardboard box. This outer box acts as a secondary layer of insulation, helping to maintain the product's frozen temperature during transportation and storage.

Both single box and box-in-box packaging methods can be effective for storing products in cold or frozen storage, depending on the specific requirements of the product and the desired level of insulation. It's essential to consider the product's temperature sensitivity, the duration of storage, and the environmental conditions to determine the most suitable packaging approach. Regular testing and monitoring of the packaging's performance in cold or frozen environments can help ensure the products remain in optimal condition until they reach the end consumers.

OBJECTIVE QUESTIONS

1. **What type of packaging is typically used for dairy products?**
   a) Glass bottles
   b) Plastic containers
   c) Metal cans
   d) Cardboard boxes
   Answer: b) Plastic containers
2. **Which packaging method is suitable for freezing products?**
   a) Vacuum packaging
   b) Cardboard boxes
   c) Glass jars
   d) Plastic bags
   Answer: a) Vacuum packaging
3. **What type of packaging is ideal for fruits and vegetables to extend shelf life?**
   a) Plastic bags
   b) Metal cans
   c) Cardboard boxes
   d) Vacuum-sealed bags
   Answer: d) Vacuum-sealed bags
4. **Which packaging material is commonly used for grains?**
   a) Glass containers
   b) Plastic bags
   c) Metal cans
   d) Paper bags
   Answer: d) Paper bags
5. **What type of packaging is recommended for pharmaceutical products?**
   a) Glass bottles
   b) Plastic bags
   c) Metal cans

d) Blister packs

Answer: d) Blister packs

6. **Which packaging material is often used for liquids like water and edible oil?**
    a) Glass bottles
    b) Plastic containers
    c) Metal cans
    d) Cardboard boxes

    Answer: b) Plastic containers
7. **What type of packaging is suitable for cosmetics and powdered articles?**
    a) Glass jars
    b) Plastic bags
    c) Metal cans
    d) Cardboard boxes

    Answer: a) Glass jars
8. **What are packaging guidelines primarily designed to ensure?**
    a) Cost-effectiveness
    b) Aesthetic appeal
    c) Product safety and integrity
    d) Environmental sustainability

    Answer: c) Product safety and integrity
9. **What is the primary advantage of single box packaging?**
    a) It is more cost-effective
    b) It provides additional protection
    c) It is more environmentally friendly
    d) It is easier to transport

    Answer: b) It provides additional protection
10. **Which type of storage is recommended for packaging that stores in cold or frozen conditions?**
    a) Dry storage
    b) Refrigerated storage
    c) Controlled atmosphere storage
    d) Humid storage

    Answer: b) Refrigerated storage

Word Limit (60-80)

1. What are the key considerations when designing packaging for special products?
2. How does packaging for cold and freezing products differ from standard packaging?
3. What packaging solutions are ideal for hot products to maintain freshness and safety?
4. What are the best practices for packaging fruits and vegetables to extend shelf life?
5. How should grains be packaged to prevent spoilage and maintain quality?
6. What specific requirements exist for packaging pharmacy products?
7. What are the key considerations for packaging liquids like water and edible oil?

Word Limit (160-180)

1. What are the key considerations for packaging special products to ensure their safety and integrity during transportation and storage?
2. How does packaging for cold and freezing products differ from packaging for products that require room temperature storage, and what are the critical factors to address?
3. What are the best practices for packaging dairy products and hot products to maintain their freshness and prevent contamination?
4. How can packaging for fruits and vegetables be optimized to extend their shelf life and minimize waste throughout the supply chain?

UNIT-3. Safety precautions

LESSON OUTLINE-

- Safety Precautions In Packaging,
- Difficulty In Packaging, Defects In Packaging,
- Label Mixing, Label Error, Lot No, Missing Lot No,
- Date Of Packaging &Manufacture
- Storage Instruction, Supervision And Examination,
- Investigation On Discrepancies And Tamper Evident.
- Packaging For Shipment,
- Packaging For Air Cargo, Rail Transport,
- Requirements Sensitivity Of Packaging For Air Cargo And
- Performance Oriented Packaging.
- Warehouse Labeling,
- Hygroscopic Protection To Be Taken Care,
- IATA Regulations And Lashing And Securing In Air Transport.

3.1 Safety precautions in Packaging

Safety precautions in packaging are essential to ensure the protection of products, the safety of consumers, and the prevention of potential hazards. Here are some key safety precautions to consider in packaging:

1. **Material Selection**: Use packaging materials that are safe, non-toxic, and suitable for the specific product.

Avoid using materials that could contaminate the product or pose health risks to consumers.

2. **Avoid Sharp Edges**: Ensure that the packaging does not have sharp edges or protrusions that could cause injuries during handling or opening.

3. **Tamper-Evident Seals**: Include tamper-evident features on the packaging to assure consumers that the product has not been opened or tampered with before purchase.

4. **Child-Resistant Packaging**: For products that could be hazardous to children, such as medications or certain household chemicals, use child-resistant packaging to reduce the risk of accidental ingestion.

5. **Correct Labeling**: Accurate and clear labeling is crucial to inform consumers about potential hazards, usage instructions, and any necessary precautions.

6. **Allergen Information**: Clearly label packaging with allergen information to alert consumers of any potential allergens present in the product.

7. **Flammability**: For products that are flammable or hazardous, ensure that the packaging materials are flame-resistant and meet relevant safety standards.

8. **Load-Bearing Capacity**: Packaging should be designed to withstand the weight of the product and any external forces during transportation and storage.

9. **Insulation for Temperature-Sensitive Products**: For products sensitive to temperature, provide appropriate insulation or temperature-controlling features to maintain product quality and safety.

10. **Hygiene and Cleanliness**: Maintain proper hygiene during the packaging process to prevent contamination of the product.

11. **Environmental Safety**: Consider the environmental impact of packaging materials and practices, opting for eco-friendly and recyclable options whenever possible.

12. **Transportation Safety**: Ensure that packaging is designed to withstand the rigors of transportation and prevent damage to the product during transit.

13. **Proper Sealing**: Adequate sealing of packaging prevents leakage, spillage, and exposure to air, which can affect product quality and safety.

14. **Storage Precautions**: Provide clear instructions on proper storage conditions for the product to maintain its integrity and safety.

15. **Quality Control**: Implement strict quality control measures to ensure that the packaging meets safety standards and does not compromise the product.

16. **Testing and Validation**: Conduct testing and validation of the packaging's performance under

various conditions to ensure it meets safety requirements.

By following these safety precautions in packaging, businesses can uphold consumer trust, minimize liability risks, and ensure that their products reach consumers in optimal condition, free from safety hazards. Additionally, compliance with safety regulations and industry standards is essential to maintain consumer confidence and product integrity.

3.2 Difficulty in Packaging

Packaging can present various challenges and difficulties for businesses and manufacturers. Some common difficulties in packaging include:

1. **Product Size and Shape**: Packaging products that come in irregular shapes or sizes can be challenging, as it requires custom packaging solutions to ensure a proper fit and protection.

2. **Fragile Items**: Packaging delicate or fragile items requires special care and cushioning to prevent breakage during transportation and handling.

3. **Cost Constraints**: Balancing the need for high-quality packaging with cost constraints can be difficult, especially for products with tight profit margins.

4. **Sustainability Demands**: Meeting consumer demands for sustainable and eco-friendly packaging

can be challenging while maintaining product integrity and shelf appeal.

5. **Regulatory Compliance**: Packaging must adhere to various regulations and standards, such as food safety regulations or hazardous materials packaging requirements.
6. **Storage Space**: Packaging materials take up space in warehouses, and optimizing storage can be a logistical challenge, especially for large quantities.
7. **Design Complexity**: Intricate or innovative packaging designs may require specialized equipment and processes, increasing production complexity.
8. **Temperature and Environmental Constraints**: Packaging for temperature-sensitive products, such as frozen or perishable items, requires careful consideration to maintain product quality.
9. **Packaging Waste**: Minimizing packaging waste and ensuring recyclability or biodegradability can be difficult, especially for complex multilayer materials.
10. **Bulk Packaging**: Packaging large quantities of products for commercial or industrial use requires durable and efficient packaging solutions.
11. **Branding and Aesthetics**: Designing packaging that aligns with branding and appeals to consumers can be a creative challenge.

12. **Global Market Compliance**: Meeting packaging requirements for international markets may involve navigating different regulations and cultural preferences.

13. **Time Constraints**: Designing and implementing packaging solutions within tight timelines can be challenging, especially for new product launches.

14. **Multi-Channel Distribution**: Packaging may need to be optimized for various distribution channels, such as retail stores, e-commerce, or direct shipments.

15. **Seasonal or Limited Editions**: Packaging designs for seasonal or limited-edition products require flexibility and adaptability.

To address these difficulties, businesses can invest in research and development to find innovative and cost-effective packaging solutions. Collaboration with packaging experts and suppliers can also provide valuable insights and expertise. Embracing sustainable packaging practices and keeping up with industry trends can help businesses stay competitive while addressing environmental concerns. Ultimately, finding the right balance between functionality, aesthetics, and regulatory compliance is crucial to overcoming packaging challenges successfully.

3.3 Defects in Packaging

Packaging defects refer to flaws or issues in the packaging that can compromise the product's quality, safety, or overall consumer experience. These defects can occur during the packaging process, transportation, or storage. Some common packaging defects include:

1. **Leakage or Spillage**: Improper sealing or damage to the packaging can lead to leakage or spillage of liquid or powdered contents, causing product wastage and potential safety hazards.
2. **Torn or Damaged Packaging**: Tears, punctures, or other damages to the packaging can expose the product to external contaminants or compromise its integrity.
3. **Incorrect Labeling**: Labeling errors, such as wrong product information, incorrect expiration dates, or missing allergen warnings, can mislead consumers and create safety concerns.
4. **Tampered Packaging**: Packaging that shows signs of tampering, such as broken seals or resealed packages, can raise doubts about product safety and authenticity.
5. **Misalignment**: Inaccurate alignment of packaging components, such as labels, sleeves, or closures, can give the product an unprofessional appearance.

6. **Inadequate Cushioning**: Insufficient cushioning or protective materials inside the packaging can result in product breakage or damage during transportation.

7. **Poor Printing Quality**: Low-quality printing, smudging, or fading on packaging labels can affect brand perception and readability of important information.

8. **Overfilled or Underfilled Packages**: Packaging that does not contain the correct amount of product as indicated can lead to customer dissatisfaction and regulatory issues.

9. **Fading or Discoloration**: Exposure to light or environmental factors can cause packaging materials and labels to fade or change color, impacting the product's aesthetics.

10. **Mold or Contamination**: Packaging that allows moisture or air to enter can lead to mold growth or spoilage of perishable products.

11. **Weak or Damaged Seals**: Inadequate sealing or weak seals can result in product contamination, spoilage, or loss of freshness.

12. **Excessive Packaging Waste**: Packaging that uses excessive materials or non-recyclable components can contribute to environmental waste and consumer backlash.

13. **Inefficient Packaging Design**: Poorly designed packaging that is difficult to open, handle, or use can create frustration for consumers.

To prevent packaging defects, companies should implement robust quality control measures throughout the packaging process, conduct regular inspections, and provide training to packaging staff. Utilizing appropriate materials, investing in modern packaging technology, and collaborating closely with packaging suppliers can also help reduce the occurrence of defects and ensure that the product reaches consumers in optimal condition.

3.4 Label Mixing, Label Error

Label mixing and label errors are two common issues that can occur during the packaging process, leading to incorrect or misleading product information on the packaging. Let's explore each of these issues:

Label Mixing:

Label mixing occurs when the wrong labels are applied to products, resulting in packaging that does not match the actual product inside. This can happen due to human error in the labeling process or issues with the labeling equipment. For example, a batch of products may be labeled with the wrong product name, ingredients, or branding, causing confusion for consumers and potential safety concerns if the product inside does not match the label.

Label Error:

1. Label errors refer to inaccuracies, omissions, or mistakes in the information printed on the labels. This can include incorrect product names, wrong ingredient lists, inaccurate nutritional information, missing allergen warnings, expiration date errors, or incorrect barcodes. Label errors can occur due to data entry mistakes, outdated label templates, or issues with the printing process.

Both label mixing and label errors can have serious consequences for both consumers and manufacturers:

- **Consumer Safety**: Incorrect labels can lead to the consumption of products that may pose health risks or trigger allergic reactions, especially if allergens are not properly disclosed.
- **Legal and Regulatory Compliance**: Labeling errors can result in non-compliance with food safety regulations, consumer protection laws, and industry standards, potentially leading to legal liabilities and penalties.
- **Brand Reputation**: Labeling mistakes can damage a brand's reputation for accuracy and reliability. Consumers may lose trust in the brand, affecting future sales and customer loyalty.

To prevent label mixing and label errors, businesses should implement rigorous quality control measures, including:

- Double-checking label information before printing and application.
- Using automated labeling equipment to reduce human error.
- Performing regular inspections and testing of labeling processes.
- Maintaining updated label templates with accurate information.
- Conducting thorough review and approval processes for label designs and content.
- Implementing proper training for employees involved in the labeling process.
- Investing in labeling technology that verifies label accuracy during application.

Additionally, having clear communication and coordination between different departments involved in the packaging process, such as production, quality control, and marketing, can help minimize the risk of label mixing and errors. Swift and transparent corrective actions are crucial if any labeling issues are detected to prevent defective products from reaching consumers.

3.5 Lot No, Missing Lot No

A "lot number" (lot no) is a unique identifier assigned to a specific batch or lot of a product during the manufacturing or production process. It serves as a crucial tool for traceability, quality control, and recall management. When a product is manufactured or processed, it is assigned a lot number, which allows manufacturers and regulatory authorities to track and monitor that particular batch throughout its entire lifecycle.

"Missing lot no" refers to a situation where a product batch or lot has been produced, but it lacks the necessary lot number identification. This can occur due to human error, equipment malfunction, or oversight during the manufacturing, packaging, or labeling process.

Missing lot numbers can lead to several issues and potential consequences:

1. **Traceability Issues**: Without lot numbers, it becomes challenging to track and trace specific product batches in case of quality issues, defects, or safety recalls.

2. **Quality Control Problems**: Lot numbers are essential for quality control purposes, enabling manufacturers to monitor the performance and consistency of products across different batches.

3. **Regulatory Non-Compliance**: Many industries, particularly in the food, pharmaceutical, and medical device sectors, have strict regulations that require products to be clearly labeled with lot numbers for traceability and recall purposes. Missing lot numbers can result in non-compliance with these regulations.

4. **Product Recalls**: In the event of a product defect or safety concern, the absence of lot numbers can make it challenging to identify and recall specific affected batches, potentially delaying the recall process and putting consumers at risk.
5. **Loss of Consumer Trust**: Missing lot numbers can erode consumer confidence in the product's quality and safety, leading to negative perceptions of the brand.

To prevent missing lot numbers, manufacturers should implement robust quality control and labeling processes. This includes:

- Implementing automation and barcode systems to reduce human errors in assigning lot numbers.
- Conducting regular inspections and audits to ensure compliance with labeling requirements.
- Verifying that each product batch is correctly labeled with its unique lot number before packaging and distribution.
- Training employees involved in the labeling process to recognize the importance of lot numbers and their proper application.

In case a missing lot number is identified, prompt corrective action should be taken to rectify the issue and ensure that accurate lot numbers are assigned to affected products. It is crucial for manufacturers to have effective systems in place for maintaining traceability and ensuring the integrity of product identification throughout the supply chain.

3.6 Date of Packaging &Manufacture

The "date of packaging" and "date of manufacture" are two important pieces of information found on product labels. These dates provide consumers with information about when the product was packaged and when it was manufactured. Let's clarify the meaning of each:

Date of Packaging:

The date of packaging is the specific date when the product was placed inside its packaging or container. It indicates the time when the product was prepared for distribution and sale. The date of packaging is particularly relevant for perishable products, such as fresh food items, where knowing when the product was packaged can help consumers assess its freshness and shelf life.

Date of Manufacture:

The date of manufacture is the specific date when the product was produced or manufactured. It represents the time when the product was created or processed, which is usually the initial step in the product's lifecycle. The date of manufacture is relevant for various products, including food, beverages, cosmetics, pharmaceuticals, and electronics, as it provides an indication of the product's age and can be used to track quality and safety.

Both the date of packaging and date of manufacture are critical for consumers to make informed decisions about the products they purchase. These dates are often printed on product labels, along with other essential information, such as expiration dates, lot numbers, and usage instructions.

For perishable products, it's crucial for consumers to pay attention to the date of packaging and date of manufacture to ensure that they are consuming fresh and safe products. On the other hand, for non-perishable items, these dates can help consumers determine the product's age and assess its shelf life or potential for use.

Regulations and industry standards may dictate how these dates should be displayed and what specific information is required on the product label. For this reason, manufacturers must adhere to labeling requirements and ensure accurate and easily readable date information is provided to consumers.

3.7 Storage Instruction

Storage instructions on product labels provide guidance to consumers on how to properly store and handle the product to maintain its quality, safety, and effectiveness. These instructions are especially important for products that have specific storage requirements to prevent spoilage, degradation, or potential hazards. Here are some common storage instructions found on product labels:

1. **Temperature**: Products may require storage at specific temperature ranges, such as "store in a cool,

dry place" or "refrigerate after opening." Temperature-sensitive items, like medications or certain food products, need to be stored within recommended temperature limits to retain their effectiveness or freshness.

2. **Avoid Direct Sunlight**: Some products, like cosmetics, medications, and certain food items, should be protected from direct sunlight to prevent degradation of ingredients and potential spoilage.
3. **Humidity**: Products sensitive to humidity, such as electronic devices, certain medications, or food items, may require storage in low-humidity environments to prevent damage or spoilage.
4. **Ventilation**: Some products may need proper ventilation to prevent the buildup of odors, moisture, or gases that could affect the product's quality.
5. **Up-right Storage**: Certain items, like chemicals, cleaning agents, or beverages, may require upright storage to prevent leakage or spillage.
6. **Dry Place**: Products like food, powders, or electronics may need to be stored in a dry place to avoid moisture-related issues.
7. **Keep Away from Children or Pets**: Products containing hazardous materials or potential choking hazards should be stored out of reach of children and pets.

8. **Avoid Freezing**: Some products, such as certain medications or liquids, should not be frozen, as freezing may alter their effectiveness or cause damage to the container.
9. **Secure Lid or Cap**: Instructions to securely close the lid or cap after use can prevent spillage and maintain the product's integrity.
10. **Shelf Life**: Indication of the product's shelf life (e.g., "use within 6 months after opening") helps consumers know how long the product can be safely used after opening.
11. **Keep in Original Packaging**: Some products need to be stored in their original packaging to maintain product freshness or prevent contamination.

It is essential for consumers to read and follow the storage instructions on product labels carefully. Proper storage ensures that the product remains safe, effective, and usable throughout its intended lifespan. Failure to adhere to storage instructions may lead to decreased product quality, safety risks, or reduced effectiveness, which can be disappointing for consumers and potentially harmful in certain cases.

3.8 Supervision And Examination

Supervision and examination are essential components of quality control and assurance processes in various industries. Both activities involve monitoring, inspecting, and assessing different aspects of products or processes to ensure compliance with standards, regulations, and desired outcomes. Let's take a closer look at each:

Supervision:

- Supervision involves overseeing and managing the activities and processes to ensure they are carried out correctly and efficiently.
- In manufacturing settings, supervisors are responsible for overseeing the production process, ensuring that workers follow standard operating procedures, and addressing any issues or deviations that may arise.
- In the context of quality control, supervisors are tasked with checking the work of employees or inspecting products to ensure they meet the required quality standards.
- Supervision also involves providing guidance, training, and support to employees to maintain high levels of productivity and quality.

Examination:

- Examination refers to a thorough inspection or assessment of products, materials, or processes to determine their quality, safety, or compliance with set criteria.
- During product examination, samples may be taken from batches or production lines and subjected to various tests to assess their

characteristics, performance, or adherence to specifications.

- Examinations may involve visual inspections, measurements, testing, and documentation of results.
- The purpose of examination is to identify any defects, deviations, or non-conformities that require corrective actions to meet quality standards.

Both supervision and examination play crucial roles in maintaining product quality and safety, ensuring compliance with regulations, and preventing defects or issues from reaching consumers. They are part of a broader quality management system that helps organizations achieve consistent and reliable results in their operations.

In the context of manufacturing and production, supervision and examination work hand in hand. Supervisors oversee the overall production process and ensure that quality control measures are implemented effectively. Meanwhile, examinations are conducted by quality control personnel or inspectors to verify that products meet the required standards.

By combining effective supervision with thorough examination processes, organizations can enhance their product quality, minimize defects and waste, and build a reputation for delivering reliable and high-quality products to their customers.

3.9 Investigation on Discrepancies And Tamper Evident

Investigating discrepancies and tamper-evident features are crucial aspects of quality control and security measures in various industries. Let's explore each of these topics:

Investigation on Discrepancies:

- Discrepancies refer to differences or inconsistencies between expected or intended results and actual outcomes.
- In the context of manufacturing and production, discrepancies can be identified during quality control checks, inspections, or audits.
- The investigation on discrepancies involves a systematic and thorough examination of the root causes of the differences to identify and address any issues that may affect product quality or process efficiency.
- Common steps in the investigation process include data analysis, reviewing standard operating procedures, conducting interviews, and examining production records.
- The goal of the investigation is to determine the reasons for the discrepancies and implement corrective actions to prevent recurrence.

Tamper-Evident Features:

- Tamper-evident features are security measures designed to show evidence of tampering or unauthorized access to a product or package.

- These features are essential for ensuring the integrity of products, particularly in industries such as pharmaceuticals, food, and electronics, where tampering can pose serious risks to consumer safety and product quality.
- Examples of tamper-evident features include seal bands, shrink bands, security labels, holographic labels, breakable caps, and tear strips.
- When tampering occurs, the features are visibly altered or broken, indicating that the product may no longer be safe for consumption or use.
- Tamper-evident features not only protect consumers but also provide a sense of security and trust in the product and brand.

Incorporating tamper-evident features helps organizations ensure product authenticity and discourage tampering, counterfeiting, or unauthorized access. For high-value or sensitive products, tamper-evident features play a vital role in maintaining security throughout the supply chain and protecting the company's reputation.

Investigating discrepancies and implementing tamper-evident features are essential components of a robust quality management system and security strategy. By promptly addressing discrepancies and ensuring product integrity through tamper-evident measures, companies can enhance customer trust, comply with industry regulations, and safeguard consumer safety. Regular monitoring, testing, and continuous improvement efforts are key to maintaining the effectiveness of these processes over time.

3.10 Packaging for shipment

Packaging for shipment is a critical step in the logistics and supply chain process. Proper packaging ensures that products are protected during transportation, reducing the risk of damage or loss while in transit. Here are some key considerations for effective packaging for shipment:

1. **Box Selection**: Choose appropriate boxes or containers that are sturdy and durable enough to withstand the rigors of transportation. The size of the box should be suitable for the product, with enough space for protective cushioning.
2. **Cushioning Materials**: Use cushioning materials, such as bubble wrap, foam inserts, air pillows, or packing peanuts, to provide a buffer and prevent items from shifting or colliding during transportation.
3. **Filling Empty Spaces**: Fill any empty spaces within the box to minimize movement and potential damage. Avoid over-packing, as it may lead to compression or distortion of the contents.
4. **Protective Wrapping**: For fragile or delicate items, consider additional protective wrapping to provide an extra layer of protection.

5. **Secure Sealing**: Ensure that the box is securely sealed using proper packing tape to prevent accidental opening during transit.

6. **Labeling and Addressing**: Clearly label the package with the recipient's address, contact information, and any special handling instructions. Include a return address for easy identification in case of delivery issues.

7. **Weight Distribution**: Properly distribute the weight within the package to maintain balance and avoid strain on one side of the box.

8. **Stackability**: If shipping multiple packages, ensure that they are stackable to optimize storage space and prevent damage during storage or handling.

9. **Waterproofing**: Use waterproof packaging materials or consider using waterproof covering to protect items from water damage during transit.

10. **Regulatory Compliance**: Ensure that the packaging complies with relevant shipping regulations and carrier requirements.

11. **Customs Documentation**: Include any necessary customs documentation for international shipments to expedite customs clearance and avoid delays.

12. **Palletization**: For larger shipments or bulk items, consider palletizing the packages for ease of handling and transportation.

13. **Tracking and Insurance**: Consider using tracking services and insurance options to monitor the shipment's progress and protect against loss or damage.

14. **Environmental Considerations**: Opt for eco-friendly packaging materials and practices whenever possible to reduce the environmental impact.

Effective packaging for shipment not only safeguards the products but also contributes to overall customer satisfaction by ensuring that items arrive in good condition and on time. By paying attention to these packaging considerations, businesses can minimize shipping-related risks and build a reliable and efficient supply chain.

3.11 Packaging For Air Cargo, Rail Transport

Packaging for air cargo and rail transport has specific considerations due to the unique challenges posed by each mode of transportation. Here are some key factors to consider when packaging for air cargo and rail transport:

Packaging for Air Cargo:

1. **Weight and Size Limitations**: Airlines often have strict weight and size limitations for cargo shipments. Ensure that the packaging complies with these restrictions to avoid additional charges or shipment refusal.

2. **Fragile Items**: Use cushioning materials, such as bubble wrap or foam, to protect fragile items from impact and vibrations during air transportation.

3. **Secure Sealing**: Properly seal the packages to prevent accidental opening during handling and transit.

4. **Documentation**: Ensure that all required shipping documents and labels are attached to the packages, including airway bills and customs forms for international shipments.

5. **Regulatory Compliance**: Packaging for air cargo must adhere to relevant aviation safety regulations, including restrictions on hazardous materials.

6. **Weather Resistance**: Consider weather conditions and possible exposure to temperature variations during transit. Use appropriate packaging to protect items from extreme temperatures.

7. **Stackability**: Design packaging that allows for efficient stacking, maximizing space utilization during air cargo transport.

8. **Visibility**: Make sure packages are clearly labeled with shipping and handling instructions for easy identification and handling.

Packaging for Rail Transport:

1. **Sturdy Construction**: Rail transportation can involve rough handling and jostling. Use durable and robust packaging to withstand the rigors of rail transit.

2. **Secure Fastening**: Ensure that packages are securely fastened and palletized when necessary to prevent movement and damage during rail transportation.

3. **Weatherproofing**: Consider potential exposure to weather elements during rail transport. Use waterproof materials to protect items from rain and moisture.

4. **Handling Labels**: Use labels indicating "Fragile," "This Side Up," or other handling instructions to guide rail personnel in proper package handling.

5. **Railcar Limitations**: Be aware of railcar dimensions and limitations to optimize space utilization and prevent package damage.

6. **Bulk Cargo**: For bulk shipments, use appropriate containers and bulk packaging methods to facilitate loading and unloading.

7. **Customs Documentation**: Include all necessary customs documentation for international rail shipments.

8. **Hazmat Compliance**: If shipping hazardous materials by rail, ensure that packaging meets regulatory requirements for safe transportation.

Both air cargo and rail transport involve handling by multiple parties, so it's crucial to design packaging that can withstand multiple handlings and potential impacts during the journey. Collaboration with shipping carriers and logistics experts can provide valuable insights into specific packaging requirements and best practices for each mode of transportation. Properly packaged cargo ensures that products reach their destinations safely and in optimal condition, minimizing the risk of damage or delays in transit.

3.12 Requirements sensitivity of Packaging for air cargo

Packaging for air cargo requires special attention and sensitivity due to the unique challenges and constraints of air transportation. The packaging must be designed to protect the contents, withstand the rigors of handling, and comply with aviation safety regulations. Here are some key requirements and considerations for packaging sensitivity in air cargo:

1. **Weight Limitations**: Airlines impose strict weight restrictions on cargo shipments. The packaging should be lightweight, yet durable enough to protect the contents without adding excessive weight that could incur additional charges.

2. **Space Utilization**: Optimizing space utilization is crucial for air cargo to maximize the payload capacity of the aircraft. Packaging should be designed to efficiently fit the cargo hold and minimize wasted space.

3. **Fragile Items**: Air cargo often involves rapid loading and unloading processes, which can subject packages to various forces. Adequate cushioning and protective materials should be used for fragile items to prevent damage during handling and transit.

4. **Stackability**: The packaging should allow for stable and secure stacking to make the best use of available space in the cargo hold.

5. **Secure Sealing**: Properly seal the packages to prevent accidental opening or tampering during handling and transit.

6. **Weather Resistance**: Air cargo may be exposed to temperature variations during transit, including extreme cold or heat. Packaging should provide some level of weather resistance to protect items from such conditions.

7. **Hazmat Compliance**: If the cargo includes hazardous materials, packaging must meet the International Air Transport Association (IATA) regulations for hazardous goods transportation.

8. **Aviation Safety Regulations**: Packaging for air cargo must adhere to the safety regulations set forth by aviation authorities, including restrictions on hazardous materials, labeling requirements, and acceptable packaging materials.

9. **Documentation and Labeling**: All required shipping documents, airway bills, and handling labels must be

properly attached to the packages for easy identification and handling.

10. **Security Features**: In some cases, sensitive or valuable cargo may require additional security measures, such as tamper-evident seals or GPS tracking.

11. **Handling Labels**: Include handling instructions on the packaging to guide personnel in proper package handling, especially for fragile or delicate items.

12. **Customs Compliance**: For international air cargo shipments, the packaging must comply with customs regulations of both the origin and destination countries.

Sensitivity in air cargo packaging is crucial to ensure that items reach their destination safely, without damage or delays. Collaborating with experienced logistics partners and understanding the specific requirements of different airlines can help businesses design packaging solutions that meet the sensitivity requirements of air cargo transportation. Properly packaged cargo not only protects the products but also contributes to efficient logistics operations and customer satisfaction.

3.13 Performance Oriented Packaging

Performance-oriented packaging refers to the design and implementation of packaging solutions that go beyond traditional protection and containment functions. It focuses on enhancing the overall performance of the product, the supply chain, and the customer experience. Performance-oriented packaging aims to improve various aspects of the packaging and delivery process, leading to greater efficiency, reduced costs, and enhanced product satisfaction. Here are some key characteristics and benefits of performance-oriented packaging:

1. **Product Integrity**: Performance-oriented packaging ensures that the product remains in optimal condition throughout its journey, reducing the risk of damage, spoilage, or contamination.

2. **Supply Chain Efficiency**: By optimizing packaging design, size, and weight, performance-oriented packaging minimizes transportation and storage costs while maximizing space utilization.

3. **Environmental Sustainability**: Packaging solutions that consider eco-friendly materials and reduce waste contribute to sustainable practices, meeting the growing demand for environmentally responsible packaging.

4. **Customizability**: Performance-oriented packaging can be tailored to specific product requirements, ensuring a perfect fit and providing superior protection during transportation and storage.

5. **Branding and Aesthetics**: Packaging that enhances product presentation and aligns with the brand image

can positively impact consumer perception and purchasing decisions.

6. **Ease of Handling**: Packaging designed for ease of handling and stacking improves efficiency during loading, unloading, and warehousing processes.
7. **Tamper-Evident and Security Features**: For sensitive or high-value products, performance-oriented packaging may include tamper-evident seals or security features to protect against unauthorized access.
8. **Enhanced Opening Experience**: User-friendly packaging that offers easy access to the product can improve customer satisfaction and loyalty.
9. **Innovation and Technology Integration**: Performance-oriented packaging can incorporate innovative features, such as smart packaging with tracking capabilities or temperature-sensitive indicators for temperature-sensitive products.
10. **Regulatory Compliance**: Packaging that adheres to industry-specific regulations and safety standards ensures product compliance and avoids potential issues during transportation and customs clearance.
11. **Multi-Channel Adaptability**: Packaging that works well across various distribution channels, including e-commerce and traditional retail, ensures a seamless customer experience.

12. **Cost Savings**: Performance-oriented packaging can lead to cost savings in terms of reduced shipping costs, minimized product returns, and improved inventory management.

Performance-oriented packaging requires a comprehensive understanding of the product, supply chain, and customer needs. It involves collaboration between packaging experts, supply chain professionals, and product designers to develop packaging solutions that align with the overall business objectives. By investing in performance-oriented packaging, businesses can gain a competitive edge, improve operational efficiency, and deliver enhanced value to their customers.

3.14 Warehouse Labeling

Warehouse labeling is a systematic and organized method of marking and identifying various locations, items, and storage areas within a warehouse. Proper warehouse labeling is crucial for efficient inventory management, accurate order fulfillment, and streamlined warehouse operations. The labeling system should be easy to understand, consistent, and scalable to accommodate the growing needs of the warehouse. Here are some key aspects of warehouse labeling:

1. **Location Labels**: Each storage location within the warehouse, such as shelves, racks, bins, and pallet positions, should be assigned a unique location label or barcode. This label provides a reference for the

exact position of products, making it easier for warehouse staff to locate and store items accurately.

2. **SKU Labels**: Every product or stock-keeping unit (SKU) in the warehouse should have a unique SKU label or barcode. This label contains essential information, including the product name, description, and identification code. It helps in accurate product tracking and inventory control.

3. **Rack and Aisle Labels**: Warehouse racking systems and aisles should be clearly labeled to guide personnel in finding specific products efficiently. Rack labels are typically placed at eye level on the rack's uprights for easy visibility.

4. **Floor Markings**: In addition to labels, floor markings, such as aisle markings and safety walkways, help maintain organized traffic flow and prevent accidents in the warehouse.

5. **Color-Coding**: Some warehouses use color-coded labels to differentiate between product categories, storage types, or picking priorities. Color-coded labels can streamline the picking process and improve warehouse organization.

6. **Barcoding and RFID**: Implementing barcode or radio-frequency identification (RFID) technology allows for quicker and more accurate data entry, inventory tracking, and order processing.

7. **Consistency and Standardization**: Ensure that the labeling system is consistent throughout the warehouse. Standardize label formats, font sizes, and placement to avoid confusion and facilitate training for new employees.
8. **Scalability**: The warehouse labeling system should be flexible and scalable to accommodate changes in inventory levels, warehouse layout, and expansion.
9. **Accessibility and Visibility**: Labels should be positioned where they are easily accessible and visible to warehouse staff. Adequate lighting and clear positioning are essential for efficient warehouse operations.
10. **Maintenance**: Regularly inspect and maintain the labels to ensure they remain readable and accurate. Damaged or faded labels should be replaced promptly.

Warehouse labeling plays a vital role in optimizing inventory control, reducing errors, and improving overall warehouse productivity. It enables employees to quickly and accurately locate products, simplifies order picking, and facilitates effective inventory management. By implementing a well-designed warehouse labeling system, businesses can enhance their logistics processes and deliver better customer service.

3.15 Hygroscopic protection to be taken care

Hygroscopic protection is essential for products that are sensitive to moisture absorption. Hygroscopic materials have a tendency to attract and absorb moisture from the surrounding environment, which can lead to degradation, spoilage, or changes in physical properties. Proper packaging and storage measures must be taken to protect hygroscopic products from moisture-related damage. Here are some considerations for hygroscopic protection:

1. **Moisture-Resistant Packaging**: Choose packaging materials that offer a high level of moisture resistance, such as foil pouches, barrier films, or vacuum-sealed containers. These materials create a protective barrier, preventing moisture from reaching the product.

2. **Desiccants**: Include moisture-absorbing desiccant packets or sachets inside the packaging to absorb any moisture that might be present. Desiccants help maintain a low-humidity environment around the product.

3. **Sealing Integrity**: Ensure that the packaging is securely sealed to prevent any moisture ingress. Proper sealing techniques, such as heat sealing or adhesive closures, are critical for effective hygroscopic protection.

4. **Storage Conditions**: Store hygroscopic products in a controlled environment with low humidity. Consider

using climate-controlled storage rooms or desiccant cabinets to maintain optimal conditions.

5. **Handling Precautions**: Handle hygroscopic products with clean, dry hands and avoid exposure to high-humidity environments during handling and packing.

6. **Avoid Condensation**: Prevent temperature fluctuations that may cause condensation on the product or its packaging. Rapid temperature changes can lead to moisture accumulation.

7. **Monitoring and Inspection**: Regularly inspect the packaging for signs of moisture ingress or damage. Implement a monitoring system to check humidity levels in the storage area.

8. **Repackaging as Needed**: If the original packaging is compromised or damaged, consider repackaging the product into new moisture-resistant containers.

9. **Transportation Considerations**: During transportation, ensure that hygroscopic products are adequately protected from rain, humidity, or water leakage.

10. **Expiration Dates**: Hygroscopic products may have a limited shelf life due to their sensitivity to moisture. Monitor expiration dates and rotate stock to ensure the use of the freshest products.

Common examples of hygroscopic products include certain pharmaceuticals, electronics, food items, chemicals, and hygroscopic powders. Proper hygroscopic protection is crucial to maintain the quality, stability, and effectiveness of these products throughout their shelf life. Businesses should consider the specific requirements of their hygroscopic items and implement suitable packaging and storage measures to preserve their integrity and performance.

3.16 IATA regulations and lashing

IATA (International Air Transport Association) regulations govern the safe transportation of goods by air, including the proper handling, packaging, and securing of cargo. Lashing is a method used to secure and restrain cargo within an aircraft's cargo hold during flight. It involves using various types of straps, nets, and other securing devices to prevent the cargo from shifting or moving during transportation. Both IATA regulations and lashing are critical aspects of air cargo safety.

IATA Regulations:

1. **Packaging Requirements**: IATA regulations specify the packaging standards that must be followed for different types of goods. These standards ensure that the cargo is properly contained and protected during handling and transit.

2. **Dangerous Goods Handling**: IATA provides guidelines for the transportation of dangerous goods by air, including the classification, packaging, labeling, and documentation requirements for hazardous materials.

3. **Load Distribution**: IATA regulations address the proper distribution of weight and balance within the aircraft's cargo hold. This is to ensure that the aircraft maintains stability and flight performance.
4. **Documentation and Labeling**: IATA mandates accurate and complete documentation for air cargo shipments, including air waybills and labels that provide information about the cargo's nature, weight, and handling requirements.
5. **Security Measures**: IATA regulations include security protocols to prevent unauthorized access to cargo and to protect against potential threats.
6. **Temperature-Sensitive Cargo**: IATA provides guidelines for the transportation of temperature-sensitive cargo, including pharmaceuticals and perishable goods, to ensure their integrity during transit.

Lashing:

Lashing is a method of securing cargo within the aircraft's cargo hold using various restraints and fastening devices. Some common lashing methods include:

1. **Cargo Nets**: These nets are used to contain and secure loose or irregularly shaped items. They prevent small items from shifting during flight.
2. **Straps and Tie-Downs**: Heavy-duty straps and tie-downs are used to secure palletized cargo or individual items to the aircraft's cargo floor.

3. **Cargo Bars**: Cargo bars, also known as load bars or cargo beams, are used to create bulkheads or barriers between cargo items to prevent movement.
4. **Pallet Nets**: Pallet nets are used to secure goods on pallets, preventing them from sliding or shifting during transport.
5. **Restraint Devices**: In some cases, specialized restraint devices may be used to secure oversized or irregularly shaped cargo.

Lashing is performed by trained personnel who are familiar with the specific requirements of the cargo being transported and the aircraft's lashing capabilities. Properly lashing cargo is crucial to prevent damage, ensure flight safety, and comply with IATA regulations.

Both IATA regulations and lashing procedures are integral to the safe and efficient transport of goods by air. Following these guidelines helps to minimize risks, ensure cargo integrity, and maintain the overall safety of air cargo operations.

3.17 Securing in air transport

Securing cargo in air transport is a critical aspect of ensuring flight safety and protecting the integrity of the cargo during transportation. Proper cargo securing methods prevent shifting, movement, or damage to the cargo during takeoff, landing, and turbulence. Here are some key considerations for securing cargo in air transport:

1. **Proper Packaging**: Cargo should be appropriately packed and secured in sturdy containers or packaging materials that can withstand the rigors of air transport.
2. **Load Distribution**: Properly distribute the weight of the cargo within the aircraft's cargo hold to maintain balance and stability. Avoid placing heavy items in a way that could cause the aircraft to become unbalanced.
3. **Palletization**: Use standardized pallets to consolidate and secure smaller items. Palletized cargo is easier to handle, load, and secure within the aircraft.
4. **Securing Devices**: Utilize suitable securing devices such as straps, tie-downs, nets, cargo bars, and load locks to prevent cargo movement. These devices should be strong and in good condition.
5. **Cargo Nets**: Cargo nets are used to contain and secure loose or irregularly shaped items. They prevent small items from shifting during flight.
6. **Tie-Downs and Straps**: Heavy-duty tie-down straps and restraints should be used to secure pallets and larger items to the aircraft's cargo floor.

7. **Securing Heavy Items**: Large or heavy items should be properly secured to the aircraft's structure or the cargo floor to prevent shifting.
8. **Secure Fastening**: Ensure that all securing devices are properly fastened, tightened, and locked to prevent unintentional loosening during transit.
9. **Padding and Cushioning**: Use appropriate padding and cushioning materials to protect the cargo from vibrations and impact during flight.
10. **Center of Gravity**: Position the cargo to maintain the aircraft's center of gravity within safe limits. This is critical for flight stability.
11. **Restrictive Access**: Restrict access to the cargo hold during flight to prevent unauthorized tampering or shifting of the cargo.
12. **Documentation and Labeling**: Clearly label the cargo with handling instructions, orientation arrows, and any special considerations for proper securing.

Properly securing cargo not only ensures flight safety but also minimizes the risk of damage to the cargo during transportation. The specific securing methods may vary depending on the type of cargo, aircraft configuration, and regulations set forth by aviation authorities. Airlines, cargo handling companies, and freight forwarders have trained personnel who are knowledgeable about the proper procedures for securing different types of cargo.

Compliance with international air transport regulations, such as those set by IATA, is essential to ensure the safe and efficient transport of cargo by air. Following established securing practices helps protect both the cargo and the aircraft, contributing to a smooth and successful air transport operation.

OBJECTIVE QUESTIONS

1. **What is the primary purpose of safety precautions in packaging?**
    a) Aesthetic appeal
    b) Product identification
    c) Protecting product and consumers
    d) Reducing packaging costs

    Answer: c) Protecting product and consumers
2. **Which of the following is considered a difficulty in packaging?**
    a) Eco-friendly materials
    b) Efficient labeling
    c) Product fragility
    d) Large packaging size

    Answer: c) Product fragility
3. **What is a common defect in packaging?**
    a) Overlapping labels
    b) Adequate protection
    c) Tamper-evident seals
    d) Barcode accuracy

    Answer: a) Overlapping labels
4. **What information is crucial on product labels?**
    a) Price only
    b) Date of packaging and manufacture
    c) Manufacturer's favorite color
    d) Marketing slogans

    Answer: b) Date of packaging and manufacture
5. **What is the purpose of storage instructions on packaging?**
    a) Decorative design
    b) Handling recommendations

c) Legal requirements
d) Brand promotion

Answer: b) Handling recommendations

6. **When should discrepancies in packaging be investigated?**
   a) Never
   b) Immediately
   c) After product expiration
   d) Annually

   Answer: b) Immediately

7. **What is the primary purpose of tamper-evident packaging?**
   a) Reducing packaging costs
   b) Enhancing product aesthetics
   c) Protecting against tampering
   d) Increasing product visibility

   Answer: c) Protecting against tampering

8. **What should be considered when packaging for shipment?**
   a) Product weight only
   b) Protection during transit
   c) Aesthetic appeal
   d) Cost of packaging materials

   Answer: b) Protection during transit

9. **What is a key consideration when packaging for air cargo?**
   a) Minimum labeling
   b) Lightweight materials
   c) Large package size
   d) Slow packaging process

   Answer: b) Lightweight materials

10. **What are IATA regulations primarily related to?**
    a) Road transportation
    b) Air transportation
    c) Maritime shipping
    d) Rail transport

    Answer: b) Air transportation

11. **What is lashing and securing primarily used for in air transport?**
    a) Labeling packages
    b) Protecting against humidity
    c) Securing cargo in the aircraft

d) Preventing tampering

Answer: c) Securing cargo in the aircraft

12. What does hygroscopic protection aim to prevent?

a) Product contamination
b) Moisture absorption
c) Label fading
d) Thermal damage

Answer: b) Moisture absorption

13. What is performance-oriented packaging focused on?

a) Aesthetic appeal
b) Cost reduction
c) Protecting the product during shipment
d) Brand promotion

Answer: c) Protecting the product during shipment

14. What is essential for effective warehouse labeling?

a) Eye-catching colors
b) Random labeling placement
c) Clear and accurate information
d) Minimal labeling

Answer: c) Clear and accurate information

Word Limit(60-80)

1. **What are the key safety precautions to consider when packaging products?**
2. **What are the common difficulties encountered in the packaging process?**
3. **What are the consequences of label mixing in packaging?**
4. **How can label errors impact product packaging and distribution?**
5. **What risks are associated with missing lot numbers on packaging?**
6. **What information should be included regarding the date of packaging and manufacture on product labels?**

Word Limit(160-180)

1. **What are the key safety precautions that should be followed in packaging to ensure product integrity during transportation?**
2. **What are the common defects in packaging, and how can they be minimized or prevented in the production process?**
3. **How can label mixing and label errors be effectively addressed to maintain accurate product information and compliance?**
4. **What are the essential requirements for packaging in air cargo, and how does it differ from packaging for other modes of transport like rail or road?**

UNIT-4. Exportable goods Packaging

LESSON OUTLINE-

- Packaging For Exportable Goods.
- Packaging For Export,
- Precaution In Packaging, Quality Packaging,
- Country's Labeling, Country's Language,
- Colour Restrictions, Provision Country Of Origin,
- ISO Standard
- Physical Protection, Quality Protection, Product Information, Machinability, Storage And
- Transport, Recyclability-Ventilation Of Package, Cushioning Materials And Moisture Absorbing System.

4.1 Introduction:

Packaging for exportable goods is a crucial aspect of international trade. Export packaging must be designed to protect the goods during transportation, handling, and storage over long distances and varied conditions. It should comply with international shipping regulations, ensure product integrity, and facilitate customs clearance. Here are some considerations for packaging exportable goods:

1. **Durable and Sturdy Materials**: Use strong and durable packaging materials that can withstand the rigors of international shipping, including handling by multiple parties, stacking, and potential exposure to adverse weather conditions.

2. **Customized Packaging**: Tailor the packaging to the specific requirements of the product, taking into account its size, weight, and fragility. Customized packaging ensures a snug fit and minimizes the risk of damage during transit.

3. **Secure Cushioning**: Provide adequate cushioning with materials like bubble wrap, foam inserts, or air pillows to protect the goods from impact and vibration during transportation.

4. **Protection from Moisture**: Ensure that the packaging offers protection against moisture, especially if the goods are sensitive to humidity. Consider using moisture-resistant materials or desiccants.

5. **Tamper-Evident Features**: Incorporate tamper-evident seals or labels to enhance the security of the goods and provide reassurance that the contents have not been tampered with during transit.

6. **Compliance with International Regulations**: Packaging for export must adhere to international shipping regulations, including those set by organizations like the International Maritime Organization (IMO) and IATA. Compliance ensures smooth customs clearance and adherence to safety standards.

7. **Labeling and Documentation**: Clearly label the packages with shipping and handling instructions, as

well as any required international symbols or markings. Ensure that all necessary export documentation, including customs forms and shipping labels, is complete and attached to the packages.

8. **Stackability and Palletization**: Design the packaging to be stackable and, if possible, consider palletization for efficient handling during loading and unloading processes.
9. **Sustainability**: Where possible, use eco-friendly and recyclable packaging materials to align with sustainable practices and meet environmental regulations.
10. **Insurance Consideration**: For high-value or sensitive goods, consider purchasing cargo insurance to protect against potential loss or damage during transit.
11. **Special Considerations**: Factor in any special handling requirements for certain goods, such as temperature-sensitive items or hazardous materials, and ensure packaging meets relevant regulations.
12. **Packaging Testing**: Conduct packaging testing to assess the durability and suitability of the packaging for the specific export conditions.

Proper packaging for exportable goods is essential to preserve the quality and condition of the products, prevent costly damage or losses, and maintain customer satisfaction.

Working closely with experienced freight forwarders or logistics providers can help ensure that the packaging meets all necessary requirements for smooth and successful export operations.

4.2 Packaging for export

Packaging for export is a critical aspect of international trade, as it ensures that goods are properly protected and transported safely to their destination. Export packaging must be designed to withstand the rigors of long-distance shipping, handling by various parties, and potential exposure to adverse conditions. Here are some essential considerations for packaging goods for export:

1. **Quality Packaging Materials**: Use high-quality and durable packaging materials suitable for the specific type of goods being exported. The materials should be able to withstand the challenges of international shipping.

2. **Customized Packaging**: Tailor the packaging to fit the size, shape, and fragility of the goods. Customized packaging reduces movement during transit and minimizes the risk of damage.

3. **Secure Cushioning**: Provide adequate cushioning using materials like bubble wrap, foam inserts, or air pillows to protect the goods from impact and vibrations during transportation.

4. **Moisture Protection**: For goods sensitive to moisture, consider using moisture-resistant packaging or including desiccants to absorb excess humidity.

5. **Proper Sealing**: Ensure that the packaging is securely sealed to prevent accidental opening during handling and transit.

6. **Tamper-Evident Features**: Incorporate tamper-evident seals or labels to enhance security and indicate if the package has been opened or tampered with.

7. **Compliance with International Regulations**: Adhere to international shipping regulations, including those related to hazardous materials, labeling, and documentation.

8. **Clear Labeling**: Clearly label the packages with shipping and handling instructions, destination addresses, and any required international symbols or markings.

9. **Documentation Preparation**: Ensure all necessary export documentation, such as commercial invoices, packing lists, and certificates of origin, is complete and accurate.

10. **Stackability and Palletization**: Design packaging to be stackable and consider palletizing goods for efficient handling and storage.

11. **Sustainability**: Whenever possible, use eco-friendly and recyclable packaging materials to align with sustainable practices.

12. **Testing and Inspection**: Conduct packaging testing to assess durability and suitability for the export journey. Regularly inspect packaging for signs of wear or damage.

13. **Insurance Consideration**: For valuable or delicate goods, consider obtaining cargo insurance to protect against potential loss or damage during transit.

14. **Special Handling Considerations**: Take into account any special handling requirements for temperature-sensitive items, hazardous materials, or perishable goods.

15. **Collaboration with Logistics Providers**: Work closely with experienced freight forwarders or logistics providers to ensure packaging meets all necessary requirements for smooth export operations.

Proper export packaging safeguards the goods, reduces the risk of damage or delays, ensures compliance with international regulations, and enhances the overall customer experience. Businesses that invest in well-designed and secure export packaging can build a reputation for reliability and maintain customer satisfaction throughout the global supply chain.

4.3 Precaution In Packaging

Precautions in packaging are essential to ensure that goods are properly protected and secure during storage, transportation, and handling. Whether for domestic use or international shipping, following these precautions can help prevent damage, spoilage, or accidents related to the packaging process. Here are some key precautions to take when packaging goods:

1. **Choose Appropriate Packaging Materials**: Select packaging materials that are suitable for the specific type of goods being packaged. Consider factors such as weight, fragility, and environmental conditions.

2. **Secure and Durable Packaging**: Use sturdy and durable packaging materials that can withstand the rigors of transportation and handling.

3. **Proper Cushioning**: Provide adequate cushioning using materials like bubble wrap, foam, or air pillows to protect the goods from impact and vibrations.

4. **Use Inner and Outer Packaging**: Consider using both inner and outer packaging to provide an additional layer of protection for fragile items.

5. **Secure Sealing**: Ensure that the packaging is securely sealed to prevent accidental opening or tampering during storage or transit.

6. **Labeling**: Clearly label the packaging with handling instructions, contents, and any special handling

requirements. Use international symbols or warnings as necessary.

7. **Weight Distribution**: Properly distribute the weight within the package to maintain balance and prevent strain on one side.

8. **Avoid Overpacking**: Avoid excessive or insufficient packing. Overpacking can lead to compression or distortion of contents, while underpacking can result in items shifting during transit.

9. **Moisture Protection**: Protect goods from moisture damage, especially for items sensitive to humidity. Use moisture-resistant packaging or include desiccants where needed.

10. **Temperature Considerations**: For temperature-sensitive items, use insulated or temperature-controlled packaging to prevent spoilage or damage.

11. **Compliance with Regulations**: Adhere to any specific packaging regulations or standards that apply to the type of goods being shipped.

12. **Customization**: Customize the packaging to fit the size, shape, and nature of the goods, especially for irregularly shaped or delicate items.

13. **Testing and Quality Control**: Conduct packaging testing to ensure the effectiveness of the packaging in protecting the goods. Implement quality

control measures to verify proper packaging standards are met.

14. **Environmental Considerations**: Where possible, use eco-friendly and recyclable packaging materials to minimize the environmental impact.

15. **Collaboration with Logistics Providers**: Work closely with logistics providers or shipping companies to understand their specific packaging requirements and guidelines.

Taking these precautions in packaging helps to ensure the goods reach their destination in optimal condition, prevent accidents, reduce losses, and contribute to customer satisfaction. Properly packaged products also help build a positive brand image and enhance the overall efficiency of the supply chain.

4.4 Quality Packaging

Quality packaging refers to packaging solutions that meet high standards of durability, functionality, and protection for the goods being packaged. It involves using materials and design techniques that ensure the integrity of the product during storage, transportation, and handling. Quality packaging plays a crucial role in safeguarding the contents, reducing the risk of damage or spoilage, and enhancing the overall customer experience. Here are some characteristics and benefits of quality packaging:

Characteristics of Quality Packaging:

1. **Durable Materials**: Quality packaging uses sturdy and robust materials that can withstand the rigors of the supply chain, including stacking, impacts, and varying environmental conditions.

2. **Secure Sealing**: Properly sealed packaging prevents accidental opening and tampering, ensuring that the contents remain intact and protected.

3. **Proper Cushioning**: Quality packaging provides sufficient cushioning and protection for fragile or delicate items, reducing the risk of damage during transit.

4. **Custom Fit**: Packaging is tailored to fit the size, shape, and nature of the goods, providing a snug and secure fit that minimizes movement during transportation.

5. **Environmental Considerations**: Quality packaging takes environmental impact into account, utilizing eco-friendly and recyclable materials whenever possible.

6. **Moisture and Weather Protection**: For goods sensitive to moisture or environmental factors, quality packaging offers adequate protection against humidity, temperature fluctuations, and adverse weather conditions.

7. **Labeling and Information**: Quality packaging includes clear labeling with handling instructions,

product information, and any necessary warning symbols.

8. **Compliance with Regulations**: Packaging meets relevant industry standards, regulations, and safety requirements to ensure proper handling and transportation.

Benefits of Quality Packaging:

1. **Product Protection**: Quality packaging safeguards the product from damage, ensuring it reaches the customer or destination in optimal condition.

2. **Brand Image**: Well-designed and high-quality packaging contributes to a positive brand image, enhancing customer perception and loyalty.

3. **Reduced Returns**: Properly packaged goods experience fewer damages, leading to reduced returns and associated costs.

4. **Efficient Supply Chain**: Quality packaging supports efficient handling, storage, and transportation, leading to smoother logistics operations.

5. **Customer Satisfaction**: Packaging that protects and presents the product well enhances customer satisfaction and confidence in the brand.

6. **Minimized Waste**: Eco-friendly packaging reduces environmental impact and contributes to sustainable practices.

7. **Compliance and Safety**: Quality packaging ensures compliance with regulations and safety standards, minimizing potential risks and liabilities.

8. **Differentiation**: Unique and well-designed packaging can help products stand out in the market and attract consumers.

Investing in quality packaging demonstrates a commitment to product quality, customer satisfaction, and responsible business practices. Businesses that prioritize quality packaging benefit from reduced costs, improved brand reputation, and enhanced overall competitiveness in the market.

4.5 Country's Labeling, Country's Language

In international trade and commerce, labeling requirements vary from country to country, and the language used on product labels is typically determined by the regulations of the importing country. Each country has its own set of labeling laws and regulations that dictate what information should be included on product labels and in what language.

Here are some key points to consider regarding country-specific labeling and language requirements:

1. Language Requirements: Many countries have specific regulations regarding the language used on product labels. In some cases, the labeling may be required to be in the official language(s) of the country where the product is sold.

This is to ensure that consumers can understand the information provided on the label.

2. **Bilingual Labeling**: In some countries, especially those with multiple official languages or a significant number of multilingual consumers, bilingual labeling may be required. This means that the label must include information in both the official language(s) and any other relevant languages.

3. **Mandatory Information**: Certain countries have specific requirements for the mandatory information that must be included on product labels. This information often includes product names, ingredients, nutritional information, safety warnings, usage instructions, country of origin, and manufacturer information.

4. **Label Format**: Some countries may have specific regulations regarding the size, font, and format of the labels. It is essential to follow these guidelines to ensure compliance.

5. **Cultural Considerations**: Businesses must be mindful of cultural sensitivities and preferences when designing product labels for different markets. Colors, images, and symbols that are considered appropriate in one country may be perceived differently in another.

6. **Labeling for Special Products**: Certain products, such as pharmaceuticals, food, beverages, and cosmetics, may have additional labeling requirements specific to their industry.

7. **Importer Responsibilities**: Importers are often responsible for ensuring that the products they import

comply with the labeling requirements of the destination country.

It is essential for exporters to thoroughly research and understand the labeling requirements of the target market before shipping their products. Working with local experts or consulting with customs agents and importers in the destination country can help ensure that labels meet all necessary legal and cultural specifications. Non-compliance with labeling regulations can lead to delays in customs clearance, fines, and potential rejection of products in the target market.

4.6 Colour Restrictions, Provision Country of Origin

Color restrictions and the provision of the country of origin on product labels are important aspects of international trade and product labeling regulations. Here's an overview of each:

Color Restrictions:

Some countries may have specific regulations or cultural preferences regarding the colors used on product labels. Certain colors may have different connotations or symbolism in different cultures, which could impact the perception of the product by consumers. In some cases, certain colors may be associated with specific product categories, and using those colors on labels could be subject to restrictions or requirements. For example, in the food industry, some countries may have restrictions on the use of artificial colors in food labeling.

It's essential for exporters to be aware of any color restrictions in the target market and ensure that their product labels comply with the local regulations and cultural norms. Working with local experts or consulting with importers or distributors in the destination country can help ensure that the label design meets all necessary requirements.

Provision of Country of Origin:

The country of origin refers to the country where the product was manufactured or produced. Many countries have regulations that require the country of origin to be clearly indicated on product labels. This information is crucial for consumers to make informed purchasing decisions and for customs authorities to determine tariff rates and import duties.

Labeling the country of origin is often a legal requirement, and failure to provide accurate information can lead to compliance issues and potential penalties. Importers and manufacturers should ensure that the country of origin is prominently displayed on the product label and complies with the regulations of the destination country.

In some cases, products may qualify for special trade agreements or preferential treatment based on their country of origin. Providing accurate and transparent information about the country of origin is essential for compliance and to maintain a trustworthy relationship with consumers and trading partners.

As labeling regulations and color preferences can vary significantly from one country to another, it's crucial for exporters to conduct thorough research on the labeling requirements of the target market. Working with local experts and trade professionals can help ensure that product labels meet all legal and cultural specifications, allowing for smooth market entry and compliance with international trade laws.

4.7 ISO standard

ISO (International Organization for Standardization) is an independent, non-governmental international organization that develops and publishes international standards. ISO standards are consensus-based documents that provide specifications, guidelines, or requirements for products, services, processes, or systems. These standards are intended to ensure safety, quality, efficiency, and compatibility in various industries and sectors worldwide. ISO standards cover a wide range of topics and are used by businesses, governments, and organizations to improve processes, enhance products, and promote best practices.

Key characteristics of ISO standards include:

1. **International Consensus**: ISO standards are developed through a consensus-based approach involving experts from various countries, industries, and stakeholders. This consensus ensures that the standards are relevant and accepted globally.

2. **Voluntary Adoption**: ISO standards are voluntary, meaning organizations can choose to implement them

as part of their practices and operations. However, many companies and industries choose to adopt ISO standards to gain a competitive advantage and demonstrate commitment to quality and best practices.

3. **Regular Review and Updates**: ISO standards are regularly reviewed and updated to reflect advancements in technology, industry trends, and best practices. This ensures that the standards remain relevant and effective over time.

4. **Broad Coverage**: ISO standards cover a wide range of areas, including quality management, environmental management, information security, occupational health and safety, food safety, energy management, and more.

5. **Consistency and Compatibility**: ISO standards are designed to be compatible with other ISO standards, promoting consistency and integration of management systems and processes within organizations.

6. **Facilitation of Trade**: ISO standards facilitate international trade by providing a common framework for products and services, reducing barriers to trade, and increasing confidence among trading partners.

7. **Global Recognition**: ISO standards are recognized and respected worldwide, making them a valuable tool for businesses operating in the global market.

Some well-known ISO standards include ISO 9001 (Quality Management Systems), ISO 14001 (Environmental Management Systems), ISO 27001 (Information Security Management Systems), ISO 45001 (Occupational Health and Safety Management Systems), and ISO 22000 (Food Safety Management Systems), among many others.

ISO standards play a crucial role in driving continuous improvement, standardization, and efficiency across industries and contribute to global economic development and sustainable practices. They provide a common language for organizations to communicate and collaborate, leading to enhanced product quality, increased customer satisfaction, and a stronger commitment to excellence.

4.8 Physical Protection, Quality Protection

Physical protection and quality protection are two important aspects of packaging that ensure the safety, integrity, and quality of the products being transported or stored. Let's explore each concept:

Physical Protection:

Physical protection in packaging refers to the measures taken to safeguard products from external factors and potential damage during storage, handling, and transportation. The primary goal is to create a barrier that shields the contents from impact, vibration, moisture, temperature variations, and other potential hazards. Key elements of physical protection in packaging include:

1. **Durable Materials**: Using strong and sturdy materials that can withstand the rigors of the supply chain and protect the contents from external forces.

2. **Cushioning and Padding**: Providing adequate cushioning with materials like bubble wrap, foam, or air pillows to absorb shocks and impacts during transit.

3. **Secure Sealing**: Ensuring proper and secure sealing of the packaging to prevent accidental openings and exposure to the elements.

4. **Moisture and Weather Protection**: Using moisture-resistant materials or incorporating desiccants to protect against humidity and environmental moisture.

5. **Temperature Protection**: Using insulated packaging or temperature-controlled solutions for goods that are sensitive to temperature variations.

6. **Impact Resistance**: Designing packaging that offers resistance to stacking and compression, reducing the risk of damage to the contents.

Quality Protection:

Quality protection in packaging is centered around preserving the integrity and characteristics of the product itself. It involves measures to maintain the quality, freshness, and safety of the goods throughout the supply chain. Key elements of quality protection in packaging include:

1. **Proper Sizing and Fit**: Ensuring that the packaging is appropriately sized to avoid compression or distortion of the contents.
2. **Product-Specific Packaging**: Using packaging tailored to the specific requirements of the product to prevent contamination or degradation.
3. **Light Protection**: Shielding light-sensitive products from exposure to UV light that can affect their quality.
4. **Airtight Packaging**: Using airtight seals to preserve the freshness and shelf life of perishable goods.
5. **Barrier Properties**: Incorporating barrier properties in packaging to protect against oxygen, moisture, or other external elements that can degrade the product.
6. **Food Safety**: Adhering to food safety regulations to prevent contamination and ensure the safety of food products.

Physical protection and quality protection go hand in hand to ensure that products are delivered to consumers in the best possible condition. By considering both aspects in the packaging design and selection process, businesses can minimize the risk of damage, maintain the quality of their products, and uphold customer satisfaction.

4.9 Product Information, Machinability

Product Information:

Product information on packaging refers to the details provided on the label or packaging that convey essential information about the product to consumers and other stakeholders. This information serves several purposes, including:

1. **Identification**: Product labels provide details like the product name, brand, and manufacturer, which help consumers identify the product.

2. **Description**: The packaging may include a brief description of the product, highlighting its features, benefits, or unique selling points.

3. **Ingredients/Contents**: For food, beverages, and personal care products, the packaging should list the ingredients or contents to inform consumers, especially those with allergies or specific dietary requirements.

4. **Instructions**: Packaging often includes instructions for product use, assembly, or maintenance. Clear instructions help consumers use the product correctly and safely.

5. **Warnings and Safety Information**: Hazardous products should have clear warnings and safety information to prevent accidents or misuse.
6. **Expiry Date**: For perishable or time-sensitive products, the packaging should display the expiry date to ensure product freshness and safety.
7. **Certifications and Labels**: Packaging may include certifications or labels indicating compliance with specific quality standards, environmental certifications, or safety certifications.
8. **Nutritional Information**: Food products often provide nutritional information such as calories, fat content, vitamins, etc., to help consumers make informed choices.

Machinability:

Machinability in packaging refers to the ease and efficiency with which the packaging material can be processed or converted using various packaging machines. It is an essential consideration for manufacturers and packaging converters as it impacts production speed, cost, and overall packaging performance. Factors affecting machinability include:

1. **Material Properties**: Packaging materials should have consistent properties to ensure smooth processing. Properties like stiffness, coefficient of friction, and static charge influence machinability.

2. **Formability**: The material should be easily formed or shaped into the desired packaging format without causing tears or deformations.
3. **Sealability**: Packaging materials should have good sealability to allow for secure seals during the packaging process.
4. **Printability**: The material should accept printing inks or labels easily without smudging or affecting the printing quality.
5. **Cuttability**: The material should be easily cut, perforated, or scored using standard packaging machines.
6. **Resistance to Abrasion**: Packaging materials should be resistant to abrasion and damage during the packaging process.
7. **Consistency**: Consistent thickness and dimensions of the packaging material ensure uniform performance during processing.

Machinability is particularly important for high-volume packaging operations where efficiency and speed are critical. Proper material selection, quality control, and testing help ensure that the chosen packaging material performs optimally in the packaging machinery, leading to smoother production and reduced downtime.

4.10 Storage And Transport

Storage and transport are integral components of the supply chain that directly impact the quality, safety, and efficiency of products as they move from manufacturers to end-users. Effective storage and transport practices are essential for preserving product integrity, reducing waste, and meeting customer expectations. Let's explore both aspects:

Storage:

Storage refers to the temporary holding of goods before they are transported or distributed to their final destination. Proper storage practices are crucial to maintaining the quality and safety of products. Here are some key considerations for storage:

1. **Optimal Conditions**: Store products in environments that maintain the required temperature, humidity, and lighting conditions. Temperature-controlled storage is critical for perishable items like food, pharmaceuticals, and certain chemicals.

2. **Segregation and Labeling**: Clearly label and segregate different products to avoid mix-ups and potential contamination.

3. **Inventory Management**: Implement an efficient inventory management system to track stock levels, expiration dates (for perishable goods), and stock rotation.

4. **Protection from Damage**: Store products on suitable shelving or pallets to protect them from damage during handling and stacking.

5. **Security Measures**: Implement security measures to prevent theft or unauthorized access to stored goods.

6. **First-in, First-out (FIFO) Principle**: Follow the FIFO principle to ensure that older stock is used or shipped first, reducing the risk of products expiring or becoming obsolete.

7. **Good Hygiene Practices**: Adhere to good hygiene practices to avoid contamination and ensure the safety of stored products, especially in industries like food and pharmaceuticals.

Transport:

Transportation involves the movement of goods from one location to another, often involving various modes of transportation such as trucks, ships, airplanes, or trains. Efficient and safe transportation is vital for timely delivery and customer satisfaction. Here are some key considerations for transport:

1. **Appropriate Packaging**: Use suitable and sturdy packaging to protect products during transit. This includes cushioning, securing, and weather-resistant materials as required.

2. **Loading and Unloading**: Properly load and unload goods to prevent damage. Use equipment like forklifts or pallet jacks to handle heavy or bulky items safely.

3. **Safety Compliance**: Ensure compliance with transportation safety regulations, such as proper handling of hazardous materials and adherence to weight limits.
4. **Route Planning**: Optimize transport routes to minimize transit time and reduce fuel consumption. Consider factors like traffic, road conditions, and tolls.
5. **Tracking and Visibility**: Implement tracking systems to monitor the movement of goods in real-time and provide visibility to customers and stakeholders.
6. **Temperature Control**: Use temperature-controlled transport for perishable goods or products sensitive to temperature variations.
7. **Proper Documentation**: Prepare accurate and complete shipping documents, including customs paperwork for international shipments.

Effective storage and transport practices contribute to overall supply chain efficiency, reduced product losses, and improved customer satisfaction. Businesses should continuously monitor and improve these aspects to ensure smooth operations and successful delivery of products to the end-users.

4.11 Recyclability-Ventilation of Package

Recyclability and ventilation of packaging are two important considerations that address environmental impact and product preservation. Let's look at each aspect:

Recyclability:

Recyclability refers to the ability of packaging materials to be collected, processed, and reused in the manufacturing of new products. Sustainable packaging design aims to reduce the environmental footprint of packaging materials by promoting recycling. Key factors related to recyclability include:

1. **Material Selection**: Choose packaging materials that are widely recyclable in the target market. Common recyclable materials include paper, cardboard, certain plastics (such as PET and HDPE), glass, and aluminum.

2. **Avoid Mixed Materials**: Minimize the use of mixed materials or composite packaging that can be challenging to recycle due to their complex composition.

3. **Clear Labeling**: Clearly label packaging materials with recycling symbols and instructions to guide consumers in proper disposal.

4. **Sustainable Printing**: Use eco-friendly printing inks and labels that do not hinder the recyclability of the packaging.

5. **Separable Components**: Design packaging with separable components to facilitate the recycling process.

6. **End-of-Life Consideration**: Consider the end-of-life of packaging materials and prioritize recyclability during the design phase.

7. **Supporting Recycling Infrastructure**: Advocate for and support local recycling infrastructure to ensure effective recycling of packaging materials.

By prioritizing recyclability in packaging design, businesses can contribute to a circular economy, reduce waste, and minimize the environmental impact of their products.

Ventilation of Package:

Ventilation of packaging is particularly relevant for products that require proper airflow to maintain their quality and extend shelf life. Inadequate ventilation can lead to spoilage, mold growth, or the build-up of unwanted odors. Considerations for ventilation in packaging include:

1. **Perforations or Air Vents**: Incorporate small holes or air vents in packaging to allow air circulation. This is commonly used in packaging for fresh produce, such as fruits and vegetables.

2. **Breathable Materials**: Use breathable packaging materials, such as certain types of paper or mesh, for products that require airflow.

3. **Perishable Goods**: For perishable goods like food or flowers, ensure that the packaging allows for proper ventilation to prevent spoilage.

4. **Humidity Control**: In some cases, ventilation may be combined with humidity control features to maintain product freshness and quality.
5. **Transport Considerations**: Account for ventilation needs during transportation to avoid airtight packaging that may lead to undesirable conditions during transit.

Proper ventilation is essential for products that are sensitive to temperature and humidity variations. It helps to preserve product freshness, extend shelf life, and maintain overall product quality, ultimately benefiting both the consumers and the brand reputation.

4.12 Cushioning Materials And Moisture Absorbing System

Cushioning materials and moisture-absorbing systems are important components of packaging, especially for protecting delicate or sensitive products during transportation and storage. Let's explore each aspect:

Cushioning Materials:

Cushioning materials are used to provide a protective layer around products to absorb shocks, vibrations, and impacts during handling and transit. They help prevent damage to the packaged items and maintain their integrity. Common cushioning materials include:

1. **Bubble Wrap**: Bubble wrap consists of plastic sheets with air-filled bubbles that create a cushioning effect.

It is lightweight, flexible, and effective in protecting fragile items.

2. **Foam Inserts**: Foam inserts or padding made from materials like polystyrene or polyurethane provide excellent shock absorption and are often custom-designed to fit the shape of the product.
3. **Air Pillows**: Air pillows, usually made from recyclable plastic, are inflated to create a cushioning barrier around products.
4. **Corrugated Inserts**: Corrugated cardboard inserts are used to add a protective layer within a package, especially for items with multiple components or delicate parts.
5. **Paper Pulp**: Biodegradable and compostable, paper pulp can be molded to fit around products, offering an eco-friendly cushioning solution.
6. **Loose Fill Materials**: Loose fill materials like packing peanuts or biodegradable starch-based pellets fill empty spaces in the package, preventing product movement and providing cushioning.
7. **Molded Pulp**: Molded pulp is created from recycled paper and molded into specific shapes to securely hold products during transit.

The choice of cushioning material depends on factors such as the fragility of the product, shipping conditions, environmental considerations, and cost-effectiveness.

Moisture Absorbing System:

Moisture absorbing systems are used to protect products from humidity and moisture-related damage, particularly in environments where moisture can negatively affect the product's quality or integrity. Moisture can lead to mold, degradation, or spoilage, especially in food, pharmaceuticals, electronics, and certain industrial products. Common moisture-absorbing solutions include:

1. **Desiccants**: Desiccants are small packets or sachets containing moisture-absorbing substances like silica gel, clay, or molecular sieves. They help maintain low humidity levels within the package.

2. **Humidity Indicator Cards**: These cards change color when exposed to specific humidity levels, providing a visual indication of moisture inside the package.

3. **Barrier Films**: Barrier films with low water vapor transmission rates can be used to create a moisture-resistant environment around the product.

4. **Vapor Corrosion Inhibitors (VCIs)**: VCIs release chemicals that prevent corrosion caused by moisture and protect metal components during transportation and storage.

5. **Absorbent Pads**: Absorbent pads are used in packaging for fresh produce and meat to soak up excess moisture and maintain product freshness.

6. **Sealed Packaging**: Properly sealed packaging prevents the entry of external moisture.

The choice of moisture-absorbing system depends on the product's sensitivity to moisture, the duration of transportation or storage, and the packaging material used.

By incorporating suitable cushioning materials and moisture-absorbing systems, businesses can ensure their products reach consumers in optimal condition, reducing product damage and maintaining customer satisfaction.

1. **What is the primary purpose of packaging for exportable goods?**
   a) Marketing
   b) Protection
   c) Branding
   d) Cost Reduction

   Answer: b) Protection

2. **Which factor is crucial in packaging for export?**
   a) Fancy design
   b) Lightweight materials
   c) Durability
   d) Colorful labels

   Answer: c) Durability

3. **What should be the top priority when packaging goods for export?**
   a) Cost-effectiveness
   b) Aesthetic appeal
   c) Quality protection
   d) Environmental sustainability

   Answer: c) Quality protection

4. **Which labeling element is important for international shipments?**

a) Brand logo
b) Country's language
c) Product description
d) Barcode

Answer: b) Country's language

5. **What does ISO stand for in packaging standards?**
    a) International Standards Organization
    b) Industrial Safety Organization
    c) International Shipping Order
    d) Integrated Sustainability Operations

    Answer: a) International Standards Organization

6. **What is the purpose of cushioning materials in packaging?**
    a) Enhance aesthetics
    b) Improve machinability
    c) Provide protection
    d) Reduce shipping costs

    Answer: c) Provide protection

7. **Why is ventilation important in certain types of packaging?**
    a) To reduce shipping weight
    b) To prevent moisture buildup
    c) To improve aesthetics
    d) To decrease recycling efforts

    Answer: b) To prevent moisture buildup

8. **What does "Country of Origin" refer to in packaging?**
    a) Where the product is sold
    b) Where the product was manufactured
    c) Where the product is currently located
    d) Where the product was designed

    Answer: b) Where the product was manufactured

9. **What is the key consideration in storage and transport packaging?**
   a) Color restrictions
   b) Moisture absorption
   c) Recyclability
   d) Product information

   Answer: b) Moisture absorption

10. **What is the primary aim of quality packaging?**
    a) Cost reduction
    b) Product protection
    c) Aesthetic appeal
    d) Easy machinability

    Answer: b) Product protection

11. **Which aspect of packaging helps reduce environmental impact?**
    a) Color restrictions
    b) Recyclability
    c) Cushioning materials
    d) Country's labeling

    Answer: b) Recyclability

12. **What should be included in the provision for country of origin?**
    a) Temperature recommendations
    b) Import duties
    c) Manufacturer details
    d) Packaging material cost

    Answer: c) Manufacturer details

13. **Which ISO standard is commonly related to packaging materials?**

a) ISO 9001
b) ISO 14001
c) ISO 22000
d) ISO 1869

Answer: a) ISO 9001

14. What aspect of packaging is essential for the machinability of goods during production?

a) Cushioning materials
b) Color restrictions
c) Labeling aesthetics
d) Size and shape

Answer: d) Size and shape

15. Which factor is NOT a concern when designing packaging for exportable goods?

a) Moisture absorption
b) Country's language
c) Recyclability
d) Marketing slogans

Answer: d) Marketing slogans

Word Limit(60-80)

1. **What are the key considerations for packaging exportable goods effectively?**
2. **How does packaging for export differ from domestic packaging?**
3. **What precautions should be taken when designing packaging for international markets?**
4. **What role does quality packaging play in ensuring the safety of products during export?**
5. **How important is labeling in compliance with the requirements of the destination country?**

6. **What factors should be considered regarding the use of colors in export packaging?**

7. **?**

8. **How do ISO standards influence the design and quality of export packaging?**

Word Limit(160-180)

1. **What are the key considerations for packaging exportable goods to ensure they meet international standards and regulations, including ISO standards?**

2. **How does proper packaging contribute to the physical protection, quality preservation, and effective transportation of exportable products?**

3. **What precautions should be taken when labeling export goods with respect to country-specific labeling requirements, language, and color restrictions?**

4. **How can packaging materials and systems be optimized for recyclability, ventilation, cushioning, and moisture absorption in order to enhance product quality and sustainability during export?**

UNIT-5. Dangerous goods Packaging

LESSON OUTLINE-

- Packaging For Dangerous Goods
- Packaging Of Dangerous Goods
- Packaging of Chemicals, Oils, Toxic Gasses, LPG, LNG.
- Can, Drum Cylinder Packaging,
- Portable Tanker Packaging, Grating Of Tanker
- Packaging Of Radioactive Materials,
- Inner Container, Outer Container And Packaging Groups I,II,III
- Pressure Receptacles For Liquids And Solids,
- Packaging Instruction Of UN3373, UN3245.
- Labelling Of Dangerous Goods, Signs /Symbols To Be Used And
- IATA Regulations For Dangerous Goods

5.1 Packaging of dangerous goods

Packaging of dangerous goods, also known as hazardous materials or hazardous goods, requires special attention and adherence to strict regulations to ensure the safe handling, transportation, and storage of these potentially harmful substances. Dangerous goods packaging is designed to contain any potential leaks, spills, or reactions that could pose risks to human health, safety, or the environment. Here are some key considerations for packaging dangerous goods:

1. **Regulatory Compliance:** Packaging of dangerous goods must comply with international, national, and local regulations specific to the type of hazardous material being transported. These regulations provide guidelines for packaging design, labeling, marking, and documentation.

2. **UN Markings and Packaging Codes:** The United Nations (UN) has established a system of markings and codes to identify different types of dangerous goods and the appropriate packaging requirements. The UN markings provide information about the nature of the hazardous material and the type of packaging required.

3. **Performance-Oriented Packaging:** Dangerous goods packaging must be designed to withstand the conditions it will encounter during transportation, including vibration, impact, stacking, and temperature fluctuations.

4. **Compatibility with Contents:** The packaging materials must be compatible with the specific hazardous substance being transported to prevent reactions or degradation that could compromise the integrity of the packaging.

5. **Leak-Proof and Secure Sealing:** Packaging for dangerous goods should be leak-proof and have secure closures to prevent any leaks, spills, or accidental openings during transportation.

6. **Absorbent Materials:** Absorbent materials may be used in the packaging to contain any potential leaks or spills and prevent them from spreading.

7. Inner Packaging and Cushioning: Some hazardous materials may require multiple layers of packaging, including inner packaging, to provide an additional level of protection and prevent breakage or leakage.

8. Venting and Pressure Relief: For certain hazardous materials that may generate pressure or gas during transportation, the packaging may include venting or pressure relief mechanisms to prevent explosions.

9. Proper Labeling and Marking: Packaging for dangerous goods should be clearly labeled with appropriate hazard warnings, handling instructions, and emergency response information. Proper markings help identify the contents and guide emergency responders in case of accidents.

10. Documentation: Complete and accurate documentation, such as shipping papers and manifests, must accompany dangerous goods shipments to provide crucial information to carriers, handlers, and emergency responders.

11. Training and Handling Procedures: Personnel involved in packaging, handling, and transporting dangerous goods should receive proper training to ensure compliance with regulations and safe practices.

Packaging of dangerous goods is a critical responsibility that requires attention to detail, compliance with regulations, and a commitment to protecting people and the environment from potential hazards. Proper packaging and handling help reduce the risk of accidents and ensure the safe movement of hazardous materials in the supply chain.

5.2 Packaging of Chemicals, Oils, Toxic Gasses, LPG,LNG.

Packaging of chemicals, oils, toxic gases, LPG (liquefied petroleum gas), and LNG (liquefied natural gas) requires specialized containers and adherence to strict safety regulations due to their hazardous nature. Let's look at the packaging considerations for each of these substances:

Chemicals:

Packaging for chemicals must be designed to prevent leaks, spills, and interactions that could pose risks to human health, the environment, and other goods during transportation and storage. Key considerations include:

- Use of UN-approved containers and packaging that meet specific performance requirements for the type of chemical being transported.
- Compatibility of packaging materials with the specific chemical to prevent reactions and degradation.
- Proper labeling with hazard warnings, chemical names, and emergency response information.
- Secure sealing and closures to prevent leakage during handling and transportation.
- Inner packaging, if required, to provide an additional barrier and prevent breakage or spills.

Oils:

Packaging for oils can vary depending on the type of oil being transported, whether it's lubricating oil, cooking oil, or hazardous oils. Key considerations include:

- Use of appropriate containers and packaging that can withstand the specific characteristics of the oil, such as temperature and viscosity.
- Leak-proof and secure closures to prevent oil spills during transportation and handling.
- Compliance with environmental regulations, especially for hazardous oils, to avoid contamination and protect the environment.

Toxic Gases:

Packaging for toxic gases involves specialized containers and cylinders designed to store and transport the gas safely. Key considerations include:

- Use of high-pressure gas cylinders made from materials that can withstand the pressure and reactivity of the gas.
- Valve protection caps to prevent accidental release of the gas.
- Proper labeling with hazard warnings and instructions for safe handling.

LPG (Liquefied Petroleum Gas) and LNG (Liquefied Natural Gas):

Packaging for LPG and LNG requires specialized tanks and containers designed to store and transport these liquefied gases safely. Key considerations include:

- Use of pressure vessels or tanks that meet safety standards and are certified for transporting liquefied gases.
- Proper venting and pressure relief systems to prevent over-pressurization of the containers.
- Leak detection systems to detect and prevent gas leaks.
- Proper labeling with hazard warnings, gas identification, and emergency response information.

For all these hazardous substances, it is essential to follow relevant international, national, and local regulations for packaging, labeling, and handling. Proper training of personnel involved in handling and transporting these materials is crucial to ensure safety and compliance with regulations. Additionally, documentation, emergency response plans, and contingency measures should be in place to address any potential accidents or incidents during transportation and storage.

5.3 Can, Drum, Cylinder Packaging

Can, drum, and cylinder packaging are commonly used for the storage and transportation of various products, including liquids, chemicals, and gases. Each type of packaging serves different purposes and is suitable for specific applications. Let's explore each one:

Cans:

Cans are small, cylindrical containers typically made of metal, such as aluminum or steel. They are commonly used for packaging beverages (e.g., soda, beer) and food items (e.g., canned fruits, vegetables, soups). Cans are known for their ability to preserve the freshness and quality of the contents, as they provide airtight and light-resistant packaging. They are also widely recyclable, making them a sustainable choice for packaging certain products.

Drums:

Drums are larger cylindrical containers made of metal (steel) or plastic. They have a capacity ranging from a few gallons to over 50 gallons. Drums are commonly used for the storage and transportation of liquids, chemicals, hazardous materials, and industrial products. They are designed to withstand the rigors of handling, stacking, and transportation and are often equipped with secure closures, such as screw caps or bungs. Drums are available in various configurations, such as open-top drums, tight-head drums, and lined drums, depending on the specific application.

Cylinders:

Cylinders are high-pressure containers used for the storage and transportation of compressed gases, such as oxygen, nitrogen, propane, acetylene, and others. These gas cylinders are typically made of steel or aluminum and have a valve for filling and releasing the gas. Cylinders are designed to withstand the pressure of the compressed gas and must meet strict safety standards to ensure safe handling and use. They are commonly used in industrial, medical, and laboratory settings.

Each type of packaging has its advantages and is chosen based on the specific requirements of the product being stored or transported. When selecting packaging, factors such as the nature of the contents, volume or quantity to be stored, transportation requirements, and safety considerations are taken into account to ensure that the packaging effectively protects the contents and complies with relevant regulations and standards.

5.4 Portable Tanker Packaging, Grating of Tanker

Portable tanker packaging refers to containers specifically designed for transporting and storing liquids or gases in bulk quantities. These tankers are mobile and often used to transport hazardous materials, chemicals, liquefied gases, and other substances that require specialized handling. Here's an overview of portable tanker packaging and the grating used in tankers:

Portable Tanker Packaging:

Portable tankers are large, cylindrical containers with reinforced walls made of steel, aluminum, or other robust materials. They are mounted on trailers or vehicles to enable transportation. These tankers come in various sizes, ranging from a few thousand liters to tens of thousands of liters, depending on the volume of the liquid or gas being transported.

Key features of portable tanker packaging include:

1. **Safety Standards:** Portable tankers must adhere to strict safety standards and regulations, particularly when carrying hazardous materials. They are equipped with safety valves, pressure relief systems, and other safety features to prevent leaks and accidents.

2. **Loading and Unloading:** Tankers are designed to facilitate easy loading and unloading of their contents, often through the use of pumps and hoses.

3. **Valves and Fittings:** Tankers are equipped with valves and fittings that allow for controlled filling, emptying, and pressure regulation.

4. **Safety Markings and Labels:** Proper labeling and markings, including UN markings for hazardous materials, are essential for identifying the contents and ensuring compliance with regulations.

5. **Insulation:** For transporting temperature-sensitive products, tankers may be insulated to maintain the desired temperature.

Grating of Tanker:

The grating of a tanker refers to the safety mesh or walkway installed on top of the tanker to provide a safe platform for workers during loading, inspection, and maintenance activities. Grating prevents workers from slipping or falling while accessing the top of the tanker.

Key characteristics of tanker grating include:

1. **Non-Slip Surface:** The grating is designed to provide a non-slip surface to ensure the safety of workers, even in wet or slippery conditions.

2. **Durability:** The grating material is chosen for its durability and resistance to corrosion, ensuring a long service life even in harsh environments.

3. **Safety Standards:** The grating must comply with safety standards to ensure it can safely support the weight of workers and any equipment used during inspections or maintenance.

4. **Easy Access:** The grating is positioned to allow easy access to the top of the tanker, with appropriate openings for loading and inspection points.

5. **Secure Attachment:** The grating is securely attached to the top of the tanker to prevent any movement or dislodging during use.

Portable tanker packaging and the safety measures implemented, such as grating, are critical for the secure transportation of hazardous substances and liquids in bulk quantities. Adherence to safety standards and proper maintenance of tankers are essential to minimize the risks associated with handling these materials.

5.5 Packaging of radioactive materials

Packaging of radioactive materials is a highly regulated and specialized process aimed at ensuring the safe containment, handling, and transportation of these hazardous substances. Packaging for radioactive materials is designed to prevent any leakage or release of radioactive substances and to protect human health, the environment, and personnel involved in the transportation process. Here are key considerations and features involved in the packaging of radioactive materials:

1. **Compliance with Regulations:** Packaging for radioactive materials must comply with international, national, and local regulations and standards specific to the type and level of radioactivity. The International Atomic Energy Agency (IAEA) provides guidelines and regulations for the safe transport of radioactive materials.

2. **Type of Radioactive Material:** The packaging design varies based on the type of radioactive material being transported. Radioactive materials are classified into different levels of radioactivity, and packaging is selected accordingly.

3. **Shielding:** For high-level radioactive materials, packaging may include shielding materials, such as lead, to reduce radiation exposure to personnel handling the package.

4. **Leak-Tightness:** Packaging must be leak-tight to prevent any release of radioactive substances during handling and transportation.

5. Criticality Safety: For fissile materials, packaging is designed to prevent criticality accidents (uncontrolled nuclear chain reactions) during transportation.

6. Containment: The packaging provides multiple barriers to contain the radioactive material, including the use of inner containers and outer containers (overpacks).

7. Radioactive Labeling: All packages containing radioactive materials are labeled with appropriate radioactive symbols and hazard warnings to identify the presence of radioactive material and inform handlers of the potential risks.

8. Security Measures: Depending on the level of radioactivity and potential security risks, packaging may have additional security features to prevent unauthorized access.

9. Impact and Environmental Resistance: Packaging must be designed to withstand impact, fire, and environmental hazards that may occur during transportation.

10. Testing and Certification: Packaging for radioactive materials undergoes rigorous testing and certification to ensure it meets the required safety standards and regulations.

11. Transport Casks: For larger quantities of radioactive materials, transport casks are used. These are heavily shielded containers designed for the safe transport of radioactive materials over long distances.

12. Emergency Response Information: Packaging must include information on emergency response procedures in case of accidents or incidents during transportation.

Packaging of radioactive materials is a critical aspect of ensuring the safe handling and transportation of these hazardous substances. It requires collaboration between regulatory authorities, packaging manufacturers, and organizations involved in the transportation of radioactive materials to maintain the highest level of safety and compliance with regulations.

5.6 Inner Container, Outer Container

Inner container and outer container are two components of packaging used for transporting hazardous or sensitive materials, including radioactive materials and other dangerous goods. These containers work together to provide a layered and robust system that ensures the safe containment and transportation of the contents. Let's explore each component:

Inner Container:

The inner container is the primary containment vessel that directly holds the hazardous material or sensitive product. It serves as the first line of defense against leaks, spills, and any potential interactions between the contents and the packaging. Key features of the inner container include:

- **Leak-Tightness:** The inner container is designed to be leak-proof and prevent any release of the hazardous material during transportation and handling.

- **Compatibility:** It is made of materials that are compatible with the specific contents to avoid any reactions or degradation that could compromise the integrity of the packaging.
- **Safety Markings:** The inner container is clearly marked with appropriate hazard warnings, symbols, and labels to identify the presence of hazardous material and inform handlers of the potential risks.
- **Criticality Safety:** For certain dangerous goods, such as radioactive materials, the inner container is designed to prevent criticality (uncontrolled nuclear chain reactions).
- **Shielding (if applicable):** In the case of radioactive materials, the inner container may have shielding materials, such as lead, to reduce radiation exposure to personnel handling the package.

Outer Container (Overpack):

The outer container, also known as the overpack, provides an additional layer of protection and containment for the inner container. It serves as a secondary barrier to contain any potential releases from the inner container and offers added durability and resistance. Key features of the outer container include:

- **Robust Construction:** The outer container is made from sturdy and durable materials, such as steel or high-density polyethylene, to withstand the rigors of transportation and handling.

- **Impact Resistance:** It may have impact-absorbing features to protect the inner container from damage during transportation, especially for sensitive or fragile contents.
- **Pressure Venting (if applicable):** The outer container may have pressure-relief valves to prevent over-pressurization due to changes in temperature or other factors.
- **Identification and Labeling:** The outer container is labeled with appropriate markings, hazard warnings, and regulatory information to ensure proper handling and transportation.

The combination of the inner container and outer container provides a secure and multi-layered packaging system that minimizes the risks associated with transporting hazardous materials or sensitive goods. This layered approach ensures that even in the event of a breach in the inner container, the outer container acts as an added containment measure, preventing any significant releases or harm. The design, manufacturing, and testing of both containers must comply with international regulations and safety standards to ensure the safe transport of dangerous goods.

5.6 Packaging groups I,II,III

Packaging groups I, II, and III are classification categories used to identify the level of danger posed by hazardous materials during transportation. These groups are defined by international regulations and help determine the type of packaging required for the safe transport of these materials. The classification is based on the hazardous properties and potential risks associated with the materials. Let's explore each packaging group:

Packaging Group I:

Packaging Group I includes substances and articles that present high levels of danger and pose significant risks during transportation. These materials have severe hazardous properties, such as high toxicity, extreme flammability, or the potential for large-scale explosions. Examples of materials in Packaging Group I include certain highly toxic substances, extremely flammable liquids, and explosives.

Packaging Group II:

Packaging Group II includes substances and articles that present a medium level of danger during transportation. These materials have moderate hazardous properties, which are less severe than those in Packaging Group I but still require careful handling and containment. Examples of materials in Packaging Group II include moderately toxic substances, flammable liquids, and certain types of corrosive materials.

Packaging Group III:

Packaging Group III includes substances and articles that present a low level of danger during transportation. These materials have relatively mild hazardous properties and pose minimal risks compared to those in Packaging Group I and II. Examples of materials in Packaging Group III include less toxic substances, non-flammable liquids, and non-corrosive materials.

The assignment of hazardous materials to their respective packaging groups is determined by various factors, including their chemical composition, physical properties, toxicity, reactivity, and potential for harm. Packaging groups are an essential component of the United Nations' system for the classification, packaging, and labeling of dangerous goods (UN Recommendations on the Transport of Dangerous Goods). These classifications help guide the selection of appropriate packaging and safety measures during the transport of hazardous materials to ensure the protection of human health, safety, and the environment.

5.7 Pressure receptacles for liquids and solids

Pressure receptacles for liquids and solids are specialized containers designed to hold and transport substances under pressure. These pressure receptacles are used for a wide range of applications, including gases, liquids, and solids that need to be stored or transported under controlled pressure conditions. Let's explore pressure receptacles for liquids and solids separately:

Pressure Receptacles for Liquids:

Pressure receptacles for liquids are containers designed to hold and transport liquids that are under pressure. These containers are commonly used for liquids that need to be stored or transported in a gaseous state or at high pressures. Some common examples include:

- **Liquefied Petroleum Gas (LPG) Cylinders:** These are portable pressure receptacles used for storing and transporting LPG, a mixture of hydrocarbon gases used as fuel in heating, cooking, and transportation.
- **Compressed Gas Cylinders for Liquids:** Some gases, such as carbon dioxide (CO_2), are transported in a liquid state at high pressure in specialized containers.
- **Cryogenic Liquid Containers:** These containers are used for transporting and storing cryogenic liquids like liquid nitrogen, oxygen, or argon, which are maintained at extremely low temperatures and high pressures.

Pressure receptacles for liquids are designed to withstand the internal pressure and temperature associated with the liquid being transported. They are typically made from high-strength materials and undergo rigorous testing and certification to ensure their safety and compliance with regulations.

Pressure Receptacles for Solids:

Pressure receptacles for solids are containers used for transporting and storing solid materials under pressure.

These containers are primarily used for materials that have the ability to generate pressure or gas when subjected to specific conditions. Examples include:

- **Pressurized Chemical Reactors:** These are vessels used for conducting chemical reactions under high pressure to control reaction rates or produce specific products.
- **Pressurized Silos:** Silos used for storing solid materials, such as grains or bulk powders, may be designed to handle pressure generated by compaction or gas generation.

Pressure receptacles for solids are designed to maintain the required pressure levels, prevent gas or pressure buildup, and ensure the integrity of the solid materials. They may include safety features such as pressure relief valves or venting systems to manage pressure and prevent over-pressurization.

Both pressure receptacles for liquids and solids are subject to stringent regulations and safety standards to prevent accidents and protect personnel, the environment, and the public during storage and transportation of hazardous materials. Proper design, manufacturing, testing, and maintenance are critical to ensuring the safe use of pressure receptacles for both liquids and solids.

5.8 Packaging instruction of UN3373, UN3245

Packaging Instruction for UN3373 - Biological Substance, Category B:

UN3373 is a code used for the classification and transport of biological substances that are known or suspected to contain infectious substances but are not in a form that poses a significant risk of infection. These substances are categorized as Category B under the United Nations' system for the transport of dangerous goods. Here are the packaging instructions for UN3373:

1. **Primary Receptacle:**
    - The biological substance must be placed in a leak-proof primary receptacle, such as a vial or test tube, that can contain any potential leaks or spills.
2. **Secondary Packaging:**
    - The primary receptacle(s) must be placed in a leak-proof secondary packaging, such as a biohazard bag or leak-proof container. This packaging ensures an additional layer of containment and prevents any leakage or contamination during transport.
3. **Absorbent Material:**
    - An absorbent material, such as absorbent pads or absorbent cushioning, must be placed between the primary receptacle(s) and the

secondary packaging. This material is used to absorb any potential leakage.

4. **Rigid Outer Packaging:**

    - The secondary packaging with the primary receptacle(s) and absorbent material must be placed in a rigid outer packaging that can withstand normal handling and prevent damage to the inner packaging.

5. **Cushioning and Padding:**

    - Sufficient cushioning material must be used to fill any voids in the outer packaging and provide protection against impact during transportation.

6. **Biohazard Marking:**

    - The outer packaging must be marked with the appropriate biohazard symbol, indicating the presence of biological substances.

7. **Documentation:**

    - Proper documentation, including the use of a shipping label that indicates UN3373 and provides relevant information about the contents, is essential for the transportation of biological substances.

Packaging Instruction for UN3245 - Genetically Modified Organisms (GMOs):

UN3245 is a code used for the classification and transport of genetically modified organisms (GMOs). The packaging instructions for UN3245 depend on the specific characteristics of the GMOs being transported and the potential risks associated with their release. The packaging instructions may vary based on the risk assessment and containment level of the GMOs.

Due to the complexity and variability of genetically modified organisms, specific packaging requirements are determined on a case-by-case basis, taking into consideration the recommendations of regulatory authorities and experts in the field of biosafety and genetic engineering.

For both UN3373 and UN3245, it is essential to comply with relevant international, national, and local regulations to ensure the safe transport of these substances while minimizing risks to human health, safety, and the environment. Proper packaging and labeling are critical to preventing accidents and ensuring that potentially hazardous materials are transported securely and responsibly.

5.9 Labelling of dangerous goods, signs /symbols to be used

Labelling of dangerous goods involves the use of signs and symbols to convey critical information about the hazards posed by the materials being transported or stored. The labelling system is standardized and follows the guidelines set by international regulatory bodies like the United Nations (UN) and the International Maritime Organization (IMO). The primary purpose of labelling dangerous goods is to provide clear and easily recognizable information to anyone handling or coming into contact with these substances. Here are the main signs and symbols used for labelling dangerous goods:

1. Hazard Pictograms: Hazard pictograms are standardized graphical symbols that represent specific types of hazards associated with dangerous goods. Each hazard class has its own unique pictogram, allowing quick identification of the primary hazard. Some common pictograms include:

- Flame: Represents flammable materials, such as flammable liquids, gases, and solids.
- Skull and Crossbones: Indicates substances that are toxic or highly toxic.
- Corrosion: Represents materials that can cause corrosion or severe damage to metals or living tissues.
- Exploding Bomb: Indicates explosives or materials that can cause an explosion.
- Gas Cylinder: Represents gases under pressure, such as compressed gases or liquefied gases.
- Environment: Indicates substances that are harmful to the environment.

2. Signal Words: Signal words are used in conjunction with hazard pictograms to indicate the level of severity of the hazard. The two main signal words used are:

- Danger: Used for more severe hazards.
- Warning: Used for less severe hazards.

3. Hazard Statements: Hazard statements provide specific information about the hazards posed by the material. They describe the nature and degree of the hazard, such as "Flammable liquid" or "Causes skin irritation."

4. Precautionary Statements: Precautionary statements offer guidance on how to handle, store, and transport the dangerous goods safely to minimize risks. They provide information on preventive measures to be taken.

5. UN Number: The UN number is a four-digit identification code assigned to each dangerous good, helping in proper identification and classification during transportation and emergency response.

6. Packaging Markings and Labels: Packaging of dangerous goods must be marked with specific labels to indicate the contents and associated hazards. These labels are usually affixed to the outer packaging and include hazard information.

7. Placards (for Bulk Transport): In bulk transportation of dangerous goods, large placards are used on vehicles to provide a visual warning of the type of hazardous material being transported.

The use of standardized signs and symbols for labelling dangerous goods is critical for promoting safety and preventing accidents during transportation, handling, and storage. It is essential for personnel and emergency responders to be familiar with these symbols to ensure appropriate precautions are taken when dealing with hazardous materials.

5.10 IATA regulations for dangerous goods

The International Air Transport Association (IATA) has established regulations for the safe transport of dangerous goods by air. These regulations are known as the IATA Dangerous Goods Regulations (DGR). The DGR provides comprehensive guidelines and requirements for shippers, airlines, and other stakeholders involved in the air transport of dangerous goods. The main goals of the IATA DGR are to ensure the safety of passengers, crew, and the aircraft, as well as to protect the environment during the transportation of hazardous materials. Here are some key aspects of the IATA regulations for dangerous goods:

1. Classification of Dangerous Goods: The IATA DGR categorizes dangerous goods into nine primary hazard classes, each represented by a specific hazard label and a UN number. The classes include explosives, gases, flammable liquids, flammable solids, oxidizing substances, toxic substances, radioactive materials, corrosive substances, and miscellaneous dangerous goods.

2. Packing and Packaging: The IATA DGR specifies detailed requirements for the packing and packaging of dangerous goods. It includes instructions for the use of appropriate UN-approved packaging, marking and labelling of packages, and the use of inner and outer packaging to ensure the safe containment of hazardous materials during air transport.

3. **Marking and Labelling:** The DGR prescribes specific hazard labels, warning labels, and markings to be displayed on packages and containers of dangerous goods. These labels and markings provide critical information about the hazards associated with the contents and assist in the proper handling and identification of dangerous goods.

4. **Documentation:** Proper documentation is essential for the air transport of dangerous goods. The IATA DGR outlines the required information to be included in the Shipper's Declaration for Dangerous Goods (DGD) and other relevant documents. The documentation must accurately describe the contents, hazards, and handling instructions for the dangerous goods.

5. **Training Requirements:** The DGR includes training requirements for personnel involved in the handling, storage, and transportation of dangerous goods. Proper training is crucial to ensure that all parties involved understand the hazards and necessary safety measures associated with the materials being transported.

6. **Restrictions and Prohibitions:** Certain dangerous goods are subject to restrictions or prohibitions for air transport. The IATA DGR lists items that are not allowed on passenger

aircraft or under specific conditions due to their hazardous nature.

7. Emergency Response: The IATA DGR includes guidelines for emergency response in the event of an incident involving dangerous goods during air transport. Proper procedures are outlined to mitigate risks and protect passengers, crew, and emergency responders.

Compliance with the IATA Dangerous Goods Regulations is mandatory for anyone involved in the air transport of dangerous goods. Airlines, shippers, freight forwarders, and handling agents must adhere to these regulations to ensure the safe and secure transportation of hazardous materials by air. The IATA DGR is updated regularly to reflect changes in regulations and to address emerging safety concerns in the transportation of dangerous goods.

OBJECTIVE QUESTIONS

1. Which packaging group is typically associated with highly hazardous materials?

a) Group I b) Group II

c) Group III d) None of the above

2. What type of packaging is commonly used for transporting toxic gases like chlorine?

a) Can b) Drum

c) Cylinder d) Portable tanker

3. What is the primary purpose of the inner container in packaging dangerous goods?

a) To provide structural support

b) To contain the dangerous substance

c) To carry additional labeling

d) To reduce packaging costs

4. Which UN number is assigned to infectious substances for transportation?

a) UN3373

b) UN3245

c) UN1203

d) UN1993

5. Which packaging instruction is associated with the transportation of biological substances?

a) PI 965

b) PI 650

c) PI 6500

d) PI 910

6. What type of packaging is suitable for transporting radioactive materials?

a) Cardboard boxes

b) Metal drums

c) Wooden crates

d) Lead-lined containers

7. What is the purpose of labeling dangerous goods during transportation?

a) To make the packaging look attractive

b) To identify the manufacturer

c) To inform handlers of potential hazards

d) To comply with local customs regulation

8. Which international organization regulates the transportation of dangerous goods by air?

a) UN

b) ICAO

c) IMO

d) WHO

9. Which of the following symbols is commonly used to indicate the presence of a flammable substance?

a) Skull and crossbones

b) Flame symbol

c) Biohazard symbol

d) Radiation symbol

10. What does LPG stand for in the context of dangerous goods packaging?

a) Liquid Pressure Gas

b) Liquefied Petroleum Gas

c) Low Pressure Gas

d) Long-lasting Gas Pressure

11. Which type of packaging is typically used for transporting corrosive chemicals?

a) Plastic bags

b) Glass bottles

c) Stainless steel drums

d) Wooden crates

12. Which packaging material is suitable for pressure receptacles used for transporting liquids and solids?

a) Plastic

b) Aluminum

c) Glass

d) Stainless steel

13. In the context of packaging for dangerous goods, what does IATA stand for?

a) International Association for Transportation Advancement

b) International Air Transport Association

c) International Association of Toxic Agents

d) International Alliance for Transportation Assessment

14. What is the primary purpose of grating on tanker packaging for dangerous goods?

a) To increase the weight of the tanker

b) To provide access for loading and unloading

c) To improve aerodynamics

d) To reduce visibility of the contents

15. Which packaging group is associated with substances presenting minor hazards?

a) Group I

b) Group II

c) Group III

d) Group IV

Answers:

1. b) Group II   2. c) Cylinder   3. b) To contain the dangerous substance

4. a) UN3373   5. b) PI 650   6. d) Lead-lined containers

7. c) To inform handlers of potential hazards   8. b) ICAO

9. b) Flame symbol   10. b) Liquefied Petroleum Gas

11. c) Stainless steel drums   12. d) Stainless steel

13. b) International Air Transport Association

14. b) To provide access for loading and unloading

15. d) Group IV

Word limit(60-80)

1. **What are the key considerations when designing packaging for dangerous goods?**
2. **How does packaging of dangerous goods differ from regular packaging practices?**
3. **What are the specific packaging requirements for chemicals, oils, toxic gases, LPG, and LNG?**

4. **What are the important guidelines for packaging radioactive materials safely?**

5. **Can you elaborate on the roles of inner containers, outer containers, and packaging groups I, II, and III in hazardous material packaging?**

6. **What are the packaging instructions for UN3373 and UN3245, and what do they entail?**

7. **What are the key provisions and regulations outlined by IATA for the transportation of dangerous goods?**

8. Word limit(160-180)

1. **What are the key regulatory requirements for packaging dangerous goods, especially those involving chemicals, oils, toxic gases, LPG, and LNG?**

2. **How do inner and outer containers, as well as packaging groups I, II, and III, impact the safe packaging and transportation of hazardous materials?**

3. **What are the latest updates and guidelines in IATA regulations for the packaging and transportation of dangerous goods, and how do they affect industry practices and safety measures?**

www.ingramcontent.com/pod-product-compliance
Lightning Source LLC
LaVergne TN
LVHW052326200926
848481LV00002B/686
* 9 7 9 8 2 2 3 4 5 7 9 7 8 *